PENGUIN BOOKS

KICKED, BITTEN, AND SCRATCHED

Amy Sutherland is the author of *Cookoff: Recipe Fever in America*, which was selected by Barnes and Noble for its Discover Great New Writers program and was included in Amazon.com's list of the best fifty books of the year. Her articles have appeared in the *Los Angeles Times, The Boston Globe*, and *Cooking Light*, among other publications. Her article "What Shamu Taught Me About a Happy Marriage" was the *New York Times*'s most e-mailed article of 2006. She was a features writer for daily newspapers in Maine and Vermont for twelve years. The author lives in Boston, Massachusetts, and Portland, Maine, with her husband, Scott, and their two dogs, Dixie and Penny.

Kicked, Bitten, and Scratched

Life and Lessons at the World's Premier School for Exotic Animal Trainers

Amy Sutherland

Penguin Books

For Ann Early, my Aunt Pretty,
who held my young fingers
to the earth's quick pulse

PENGUIN BOOKS
Published by the Penguin Group
Penguin Group (USA) Inc., 375 Hudson Street, New York, New York 10014, U.S.A.
Penguin Group (Canada), 90 Eglinton Avenue East, Suite 700, Toronto,
Ontario, Canada M4P 2Y3 (a division of Pearson Penguin Canada Inc.)
Penguin Books Ltd, 80 Strand, London WC2R 0RL, England
Penguin Ireland, 25 St Stephen's Green, Dublin 2, Ireland (a division of Penguin Books Ltd)
Penguin Group (Australia), 250 Camberwell Road, Camberwell,
Victoria 3124, Australia (a division of Pearson Australia Group Pty Ltd)
Penguin Books India Pvt Ltd, 11 Community Centre,
Panchsheel Park, New Delhi – 110 017, India
Penguin Group (NZ), 67 Apollo Drive, Rosedale, North Shore 0745,
Auckland, New Zealand (a division of Pearson New Zealand Ltd)
Penguin Books (South Africa) (Pty) Ltd, 24 Sturdee Avenue,
Rosebank, Johannesburg 2196, South Africa

Penguin Books Ltd, Registered Offices:
80 Strand, London WC2R 0RL, England

First published in the United States of America by Viking Penguin,
a member of Penguin Group (USA) Inc. 2006 Published in Penguin Books 2007

10 9 8 7 6 5 4 3 2 1

THE LIBRARY OF CONGRESS HAS CATALOGED THE HARDCOVER EDITION AS FOLLOWS:
Sutherland, Amy.
Kicked, bitten, and scratched : life and lessons at the world's premier
school for exotic animal trainers / Amy Sutherland.
p. cm.
ISBN 0-670-03768-0 (hc.)
ISBN 978-0-14-311194-8 (pbk.)
1. Moorpark College. Exotic Animal Training and Management Program
2. Animal trainers—Vocational guidance—California. 3. Exotic animals—California.
I. Title.
SF83.C23M667 2006
636.088'8023—dc22 2005057474

Printed in the United States of America
Designed by Nancy Resnick

Contents

Preface

While working on this book, spending hours watching animal trainers accomplish the seemingly impossible, teaching a hyena to fetch and a baboon to ride a skateboard, I got to thinking of a certain species—the American husband. Maybe, I thought, the techniques used to teach elephants to paint and dolphins to flip, might work with this animal, in particular with the subspecies known as Scott Sutherland.

At home, I went to work. I adopted the trainer's motto "the animal is never wrong." I analyzed my husband's behavior like a trainer does a badger's or a tiger's. I used techniques with scientific names. I did this half as a goof—nothing I tried before was this much fun—but in short order I got results.

On June 25, 2006, I announced via a column in *The New York Times,* how I had improved my marriage using the principles of progressive exotic animal training. By noon or sooner, my column, entitled, "What Shamu Taught Me about a Happy Marriage," became the most e-mailed story off the *Times* Web site. There it remained for the next thirty days, much to my shock.

My phone began to ring. I was interviewed by reporters in Australia, Colombia, Brazil, France, Belgium, Turkey, and Germany. My column was blogged. I was called a devil shrew on a right-wing Web site, which I consider a kind of honor. I was interviewed on NPR, MSNBC, and the *Today Show*. My inbox filled up with fan mail. A few psychiatrists wrote to congratulate me. An acquaintance was given a copy of my column by her marriage therapist. My musings inspired a *Sylvia* cartoon.

Before long, I had a new book contract followed by a movie deal. Now exotic animal training had not only improved my marriage but changed my life.

I was asked over and over why I thought my column had hit such a chord. The short answer is, I really don't know. The long answer is I think people are hungry for a fresh way to negotiate the small, daily annoyances that can poison a marriage. You don't want to go to a counselor for every little thing, but all those little things in sum can finally land you there. Being social animals, we instinctively want to get along, especially with our spouses. But as another sock hits the floor or toilet seat is left up, the fur flies.

Second, I think many people, like myself, welcome proof that, indeed, we humans are squarely in the animal kingdom. There's something comforting to knowing that the same approach that can be used to teach a hermit crab to ring a bell for its dinner or a killer whale to breach on command can be used to teach us highest of primates to take the trash out. We may have the biggest, wrinkliest brains, but that does not make us immune to basic behavioral principles that apply to all organisms. We are truly not alone on this planet.

Nearly four years ago, I walked into an unusual zoo in southern California never, ever expecting that I was embarking on a path that would lead far beyond the compound's high fence, both figuratively and literally.

Trainers teach what they call an "A to B," meaning getting an animal to go from here to there. Writing and reporting has trained me to go from A to wherever. B is always out there, somewhere, but it's never where or what I expected. Thank god.

Introduction

I have always considered myself an animal person, meaning I not only felt at ease with most things furred, feathered, scaled, even fanged, but moreover, I found them an endlessly compelling and integral part of the world. A life without this passion has always struck me as a lesser one, and I admit to being a hard judge of people who do not share my love for the animal kingdom. At worst, I consider them suspicious; at best, deeply flawed. Perhaps my love of animals, which I admit I consider a superior trait, is simply a by-product of my upbringing. Despite its standard white suburban backdrop, my childhood managed to stray from the stereotype, thanks to a fun-loving mother and the unruly woods that breached our trim backyard.

I grew up with the requisite cats and dogs, mostly strays. There was Curly, a handsome mutt with a wavy coat worthy of a Breck Girl. There was Tang, a low-to-the-ground terrier mix, who always ran in a diagonal line and whose sizable balls, which nearly dragged between his short legs, could not go unnoticed even by the most modest of dog lovers. We also had three ducks one summer that chased us while we played Wiffle ball and overfertilized our and our neighbors' yards with their voluminous and slippery green dung. We temporarily housed turtles that lumbered along the road after rainstorms. One had three legs and dragged a corner of its shell in a way that broke our hearts. We kept it longer than the requisite week and fed it full of iceberg lettuce. One Easter we got a black-and-white bunny that hopped around the house, despite our cat and dog, depositing neat piles of raisin-sized poops my mom vacuumed up. When the bunny became a good-sized rabbit and the poops grew plumper, we put it outside, assuming it would prefer to live in the great outdoors with other rabbits. Instead, the bunny stayed close to home, hopping among the shrubs in our front yard. Most nights, the rabbit would wait in the garage for my father to come home from work and scratch its long ears. When I was in high school we got a dove, Eleanor, who spent her days cooing to her image in a mirror, laying eggs in our hanging plants, and trying to land on our heads. There were also guppies and hamsters along the way, though we were put off by their maternal cannibalism. The sound of a mother hamster munching her baby with lip-smacking relish still rings in my ears.

Beyond our household and its menagerie, we found the animal kingdom thriving in our suburban neighborhood. Just a few houses down, an older girl I much admired kept pet chickens that ran loose, lending an incongruous country touch to the new suburb. A couple of streets over, a raucous family of nine kept baby raccoons. My sister, our friend Bunny, and I spent whole afternoons wading ankle-deep in a wide creek, turning rocks over, looking for crayfish. The few we managed to get our small hands on—they were fast—we'd collect in a pail, only to empty it back into the creek's shallow, clear waters before heading home for cream-cheese-and-jelly sandwiches. At night, we tossed pebbles into the dark sky, provoking bats to dive-bomb us as we threw ourselves flat on the grass. We hiked through local farm fields filled with cows without a second thought, until one chased us down a hill. We only escaped being stampeded by throwing ourselves over a fence; then we lay there windless and laughing. The marauding cow stood stolidly by on the other side of the fence staring at us, as if to make its point.

When I left home and took up apartment life, my love of animals was mostly expressed vicariously. None of the landlords allowed dogs. I am allergic to cats. I would never, ever get another hamster. Besides, my life was too peripatetic for pets. Urban animals were a touch too urban, such as the obese raccoon that each night climbed a fire escape on my building, begging for food from window to window. Squirrels ate a hole in the kitchen screen of one apartment and regularly slipped in while I was out, raiding my trash can or pilfering a muffin I'd left on the counter and then, for some reason, retiring to my bathtub for their feast.

I was left to satisfy myself by petting any dog that came within arm's length and watching all things nature on television. I dreamed of buying a house, not to have the house so much as to get a dog, to have a small yard where I could hang a bird feeder, and maybe even to find a snake or two in the grass. Finally, in the spring of 1999, having acquired a house and a bit of stability, my husband and I drove two hours down the coast of Maine to New Hampshire to the home of a haughty breeder of Australian shepherds. We were there to collect an eight-week-old female red tricolor. As we expected, we fell hard. After being sized up by the taciturn breeder and deemed worthy—just barely—of an Aussie, she lent us a puppy crate, and we loaded our furry charge into the backseat. On the drive home, the pup whimpered in her crate on the backseat as we debated names and unwittingly drove through the night to a new and much richer life.

I had grown up in the era when everybody let their suburban pooches run. I had neither trained a dog nor known a trained dog. Dixie, as we named her, required training. She's a herding dog, and herding dogs need direction, as trainers say euphemistically. So Dixie and I went off to puppy school and then to general obedience and agility classes, all taught by an enlightened trainer I happened upon. To my surprise, I did most of the learning and struggled to keep up with Dixie, a far quicker study. The classes, what with the barking and trying to hold the leash, clicker, and treats in one hand, could be unbelievably aggravating. I felt like a miserable failure most of the time. I stuck with it, because it soon became clear that training opened a two-way line of communication between me and Dixie. More than that, it gave me a way to communicate with another species—an unexpected thrill. I felt an underused lobe in my brain rev up each time I worked with Dixie. It was a blissfully nonverbal lobe that read body language, one that considered the world from another species' point of view. Briefly, I was liberated from my overwhelming humanness. I was hooked.

Eventually, I found myself on the Paris set of *102 Dalmatians* on an assignment for *Disney* magazine. There was a small army of trainers with their menagerie of animals, mostly dogs (a borzoi, two French poodles, a bulldog, a border terrier, and, of course, dalmatians), and a few parrots. I hung out with the trainers from Birds & Animals Unlimited day after day, watching how they calmly got their dogs to run through moving traffic or how they hid behind trees or mailboxes to give their canine actors hand commands during takes. If I wasn't with them, I was chewing the fat with the American Humane Association rep on the set, who was a horse trainer also. I basically talked animals all day. I talked about how to knit a sweater for a dachshund, why a French trainer's authoritarian style wasn't working with his two poodles, and how to tell when a dog was stressed. On the film set, I was surrounded by the nerve-wracking, nitty-gritty construction of fiction, but I had taken refuge in a real world as magical as make-believe, a world in which people knew essentially how to talk to animals. In my notes was the name of the school nearly all the trainers had attended: a small community college in Southern California called the Exotic Animal Training and Management Program at Moorpark College. That such a place, a college that taught people to train everything from rabbits to tigers, existed struck me as incredible. When I got home I wrote MOORPARK on a piece of lined paper. Inspired, I taught Dixie to bring in the newspaper.

Our passions often lead us down paths that we don't even realize we are on until we arrive at an unexpected but welcome destination. In July 2003, I found myself standing atop an arid hill on the outside of a high chain-link fence with a gregarious vet by my side. A sign read America's Teaching Zoo. That turned out to be a humble claim for what lay beyond the gate, as my guide, the vet, would show me. Within was a secret garden full of small mysteries, deep passions, unexpected heartbreak, rare beauty, and high comedy. For a year I lived within what seemed at times to be a childhood storybook come to life. I went on emu walks, smiled at a baboon who smiled back, got a kiss from a sea lion, played with a baby orangutan, and touched my first snake—a king snake. I watched as students taught a hyena to spin, a camel to play basketball, and a baboon to get in a crate and close the door behind itself. I spent long days among fellow animal lovers to whom I never had to explain why I so adored the pygmy goat or why I never tired of going for walks with the cougar brothers. In a way, I suppose, I was reliving those aimless days of my youth that were filled with crayfish, bats, frogs, dogs, and chickens. Only now they were full of badgers, servals, miniature horses, and capuchin monkeys.

I had come cross-country for a good subject. What I didn't expect was that my time at the teaching zoo would change me, corny as that sounds, and for the better, as I suspect it does anyone who spends much time there. For who can spend any time within a magical kingdom, even a flawed one, and not return to the black-and-white world changed, if only by knowing that the seemingly impossible does exist?

Orientation

On an August day, under a sharp blue California sky with a view of the umber Santa Susana Mountains behind him so beautiful it can make you forget the pounding 100° heat, Dr. Jim Peddie stands in the shade and speaks of death. As a veterinarian who has euthanized hundreds of people's beloved pets during his long career, he knows death too well but he has never grown comfortable with that moment when life slips away at his say. "Everyone thinks death should be peaceful, but it seldom is," he says, his hands in his jeans pockets, his face pinched, and his voice raw.

Before him are fifty-one faces scattered over metal risers in a small outdoor theater. The smooth, tan faces belong to the incoming class of students or—as they will be referred to for the next twelve months—the *first years* in the Exotic Animal Training and Management Program at Moorpark College (EATM). This new crop of aspiring exotic animal trainers are nearly all women, forty-seven out of fifty-one. Most are in their early to mid-twenties, many of them tall. They are dressed for the heat in shorts, visors, and tank tops.

Tattoos scroll across their shoulders or lower backs. They look eager, optimistic. This is their first step toward a bright, sunny future. Death—that dark, distant star—is the last thing on their minds.

Still, Dr. Peddie's gravity is not lost on them. Nobody smiles. Their sunglassed eyes all rest quietly and attentively on the broad-shouldered, fatherly vet. The change in tone is oddly striking in what has been up to now an overwhelming, yet giddy, few days of meet and greets. On their orientation week schedules, this one-hour slot is listed blandly as Processing Food Animals. Most of the new first years know what is coming and have steeled themselves, though there's a rumor they may be spared this gruesome initiation rite that requires animal lovers to prove their love by killing a bird with their bare hands. It's an early litmus test of whether the first years are tough enough for the program, because the school is not, as Dr. Peddie says, for people who think animals are cute.

Birds of prey and reptiles require freshly killed prey, Dr. Peddie explains. In captivity, some can be trained to eat dead animals, but others must have their caretakers hunt for them. Consequently, the school teaches students how to humanely kill pigeons and rats. Every student must break a pigeon's neck with her hands, what they call pulling a pigeon, or gas a rat before she can graduate. There is no way around it, the vet explains. Crying vegetarian won't get you out of it, nor will your religious beliefs. "I feel it is important you do it so you know you can do it," he says. "We've had [graduates] lose jobs because of this. You people are animal people, and this is part of animal care. We do this right up front and early."

He describes how the birds' wings flutter, the small black eyes blink, and the head pops off in your palm. As you pull, you may feel the spinal column stretch like a piece of elastic. Despite the medieval style of execution, this is the quickest way to render the birds unconscious, he says, and is thus the most humane. "People deal with this differently," he explains. "Some people will break down crying, some will burst out laughing like they are giddy. Either one is the same thing, a release, so don't be critical of how someone reacts. They're not laughing because they are ecstatic. They are ecstatically uncomfortable."

Usually at this point, a current student gently takes a bird, wings flapping, with one hand and leans over a trash can. With her other hand, she quickly jerks its nut-shaped head, cracking its small vertebrae and tearing the neck; she drops the head, which lands with a small thump on the bottom of the can. Then Dr. Peddie asks for a half dozen volunteers to step forward for this odd baptism. Today, this is not to be. After all this buildup, it turns out the rumor is true. Fortunately or unfortunately (depending on whether you'd just as soon get it over with or prefer to procrastinate pulling a pigeon until the very last day of school), there are no birds to kill. To head off the spread of Newcastle disease, a contagious and deadly virus, a statewide quarantine has stopped all sales of birds, Dr. Peddie explains. He hopes to have some pigeons soon, he says. He apologizes.

Death, however, will not be denied on this sunny day. Instead, a half dozen unlucky rats will be smothered with CO_2. Before a collective sigh of relief can be exhaled, a trio of women in their twenties wheels in a shoulder-high metal tank of the gas, precariously strapped to a handcart by a bungee cord, and totes in a plastic bin of rats. These women are *second years,* meaning they are in the second year of the program. Moreover, they are the Rat Room managers. They oversee a small colony of rats in a room the size of a large walk-in closet. The rats are raised to feed the zoo's reptiles and birds of prey. A Rat Room manager with schoolmarm glasses and visor pulled low cinches a spotted rodent around the shoulders with her thumb and index finger and hoists it up for all to see. Except for its busy nose, the bright-eyed rat goes slack, its pink tail hanging straight. "Be careful, because they will, I repeat *will,* bite you," she says.

Everything is ready to go. All that is needed is a plastic bin, a small plastic garbage bag, the CO_2, and, of course, the rats. The managers are at pains to explain themselves and keep repeating that they do not relish their task. "Then we feed the golden eagle and see the enjoyment he gets out of the food. It's the circle of life," one manager, unsmiling and squinting in the sun, explains. "If you are upset by it, if you want to cry, go for it," she says. "If it's upsetting to you, let your emotions out."

A Rat Room manager quickly, unceremoniously loads six rats, noses twitching, little ears upright, into a plastic bin covered by a small green garbage bag. There's hardly enough time to spit out a good-bye. Another manager closes the bag around the carbon dioxide tube. The other holds her hands down on the rats, because they sometimes push their way out. The third manager opens the gas valve. In the bleachers, no one says a word. The two minutes tick away slowly as everyone stares at the plastic bag. There are no noticeable rustlings in the bin. No squeaks for help. The gas is turned off. The now limp rats are removed one by one. The managers lightly tap the eyes with their index fingers to make sure the rats are stone dead. There is no blood, no smell.

While men are embarrassed to cry, women can be embarrassed not to, but there is nary a sniffle. Instead of a wet-hanky fest, there is a solemn hush. This is broken as the new first years raise their hands and ask practical questions like how often do they gas rats and for how long exactly. The bin of gassed rats is whisked away to a freezer. The canister of CO_2 is wheeled offstage. That's enough of death for one day.

A sheep zips across the back of the outdoor stage. A pig, his hide a sooty black, ambles out and pokes at a ratty red carpet with his nose until the length of it unfurls and two chunks of apple pop out. Having devoured them, he lazily saunters offstage, his scrawny tail giving a little twitch as he exits. "We're going to need to cut his tusks again," Dr. Peddie sighs, sitting in the bleachers next to me. The first years will find that the sudden shift in mood is emblematic of life at EATM, where emotions run high and the unexpected is around every corner.

Orientation is packed with traditions, one of which is the off-color show the second years present. It's a chance for them to strut their training stuff and cut loose after a long, grinding summer of running the teaching zoo by themselves. What follows is a ribald beauty pageant of beasts that breaks all the rules of a proper animal education presentation. They even have the animals do *tricks*, a forbidden word among enlightened trainers, who prefer *behaviors* instead. In one hand, the MC carries two dead squirrels frozen in an amorous embrace. He occasionally holds them to his sweaty brow. When a student rides Kaleb, the caramel-colored camel, onstage, the MC says, "Here's a big hairy beast onstage with a camel underneath." He warns that Kaleb could "freak out at any minute" and notes that a camel has thick knee pads and prehensile lips. "When would that come in handy?"

Another student totes Happy, the American alligator, onstage like a big log. The MC rattles off some alligator stats—they are exothermic, only grow as big as their enclosure, and have extra eyelids—then encourages his audience to take the Velcro strap off his snout. "Really, it's like opening a present." He adds, "Their skin makes excellent shoes and purses." He pauses. "Would you ever say that in a regular show?" None of the first years answers. Just laughter. "First years, what have you learned in your first three days? Wake up!" he taunts.

Half the show's humor comes at the student handlers' expense. A good number of the animals don't do as told. A little big-eared fox suddenly bounds off a student's shoulder as she exits the stage; despite her trainer's protests, C.J., the coyote, takes a long drink of water from a shallow moat that rings the stage; Julietta, the emu, won't take her exit; Banjo, the macaw, won't get on his roller skates.

Finally, the stage is hosed down, techno music is cranked up, and the star arrives. Schmoo, the twenty-four-year-old sea lion, head up and barking joyfully, bounds onstage like a rock star with her band—in this case, her four student trainers. Schmoo isn't just the star of the show but of the whole program, with her 170-plus commands and her long list of movie and commercial credits.

Schmoo zips back and forth between the student trainers as they put her through her paces, tossing her chunks of slippery squid. She quickly rolls over and coats herself with specks of dirt, barks jubilantly, raises a flipper to her brow in a quick salute, and sticks her tongue out like a third grader. When a trainer points her finger and says, "Bang!" Schmoo collapses in an overly dramatic heap worthy of a silent screen diva. When a trainer says, "Shark!" she tosses a flipper up to imitate the killer. She tips far forward on her breast and pitches her tail happily into the air. Then Schmoo does the reverse, rising up on her tail, throwing her flippers out, and pointing her nose heavenward like an angel.

The first-year students are rapt. They lean forward and smile broadly. This is why they are here, why they'll endure a brutal schedule, give up their social life, and take on huge student loans. A year from now it could be they who are having a high time tossing squid to this incredible creature and singing out "Shark!" This is proof, however fleeting, that their farfetched dreams of working with animals can come true. What they don't know is that, backstage, Gabby, the Catalina macaw, has bitten one of the second years badly enough that she has been rushed to the campus Health Center. Dreams always come with a price, whether it's money, time, or blood.

It is now the third day of what is likely to be the hardest twenty-one months of any first year's life. This orientation week—a busy string of potlucks, ice breakers, and gag gifts—is a deceptive introduction to life at the school. However, the week is peppered with advice and announcements that foreshadow what's ahead. Starting next week, the first years won't have an official vacation until next summer. They will work most, if not all, holidays and most weekends. Four days a week they are due here by 6:30 A.M. and won't leave until 5 P.M. During these long days, they will care for the teaching zoo of some 150 to 200 animals, doing everything from hosing out the cages to answering the phones. When not cleaning, feeding, or even weeding, they will attend classes—one of the few times they get to sit down during the day, which often induces deep naps complete with drooling and snoring. In the evening, drowsy from the day, they will study animal anatomy tomes and memorize agonizingly long lists of Latin species names. As an alum puts it, the school "pretty much owns you."

Students will follow a list of rules worthy of the Marines. Their uniforms must be kept clean; shorts cannot be too short; earrings cannot be too long; bra straps cannot show; sleeves must not be rolled. They cannot use their cell phones at the zoo. They cannot smoke. They cannot run or laugh near the primates' cages. Most important, they cannot be late for the morning cleaning. If these rules are repeatedly broached or if a grade slips below a C in any class, they will be kicked out. Four students among the class of 2004 got the boot last year.

They could be chomped, mauled, or even killed by an animal. Even the smallest nick could produce a surly infection. They might catch a zoonotic disease—anything from parrot fever to bubonic plague. "Wash your hands," a staffer reminds them. "The last thing we want is any of our students coming down with worms." Dr. Peddie warns that their romances, especially new ones, may not survive. They will not have a social life beyond the zoo. Tell your family you'll hardly see them, Dr. Peddie instructs. Working part-time is strongly discouraged. Money will be tight. As the letter to prospective students ominously warns, "You will not have much time for yourself, so make sure areas of your life are in order."

They will live and die by their planners, in which they will chart out their days in fifteen-minute increments and write endless to-do lists. They will get what they call EATM hands: hands that are cracked, cracks that are filled with dirt, dirt that can't be washed out, no matter how hard they scrub. They will keep an extra set of clothes or two in the car because, as a staffer tells them, "If the tiger sprays you, you don't want to go around smelling like urine all day."

The students will suck it up and count their blessings because they have gotten into the premier program in the country, if not the world. This is the Harvard for exotic animal trainers. It's also the only academic program for trainers. Santa Fe Community College in Gainesville, Florida, has a zoo as well, but its program is primarily for zookeepers. If you want to be a trainer, Moorpark is the school for you. The general public may never have heard of it, but anyone in the animal industry has. Graduates work in zoos across the country, in Hollywood, the U.S. Navy, Ringling Bros., Guide Dogs for the Blind, and SeaWorld. They work in sanctuaries, aquariums, animal parks, and research facilities. Most of the trainers at Universal Studios' animal show are Moorparkers, as they are known. A few years back, the San Diego Zoo and Wild Animal Park, one of the country's top zoos, filled seven openings with recent grads of the program. Julie Scardina, the animal ambassador for SeaWorld and one of the country's most visible trainers with regular stints on *The Tonight Show* with Jay Leno, is a Moorparker.

EATM perches on a ridge overlooking the college's ultragreen football field and driving range on one side, the orderly houses of the commuter burg on the other side. To the east is a range of arid, muscle-bound mountains. To the west, the four lanes of a highway bend through a treeless landscape. The highway occasionally backs up, reminding you that Los Angeles is only fifty miles to the southwest.

EATM's entrance is tucked into an out-of-the-way corner of the community college, at the far end of its moatlike parking lot. The sign out front reads America's Teaching Zoo, but no one calls it that except when they answer the telephone. Students and staff refer to the program as EATM (pronounced "EE-tem") or just Moorpark. Looking through the high chain-link front gate, all you can see is a small outdoor theater, an unassuming huddle of buildings, some olive trees, and a single-lane blacktop road running away from you. There is not an animal in sight, though you might hear a doleful wolf howl or a cranky cockatoo shriek. Those are the only clues to the reason why this gate must always be kept closed.

Just inside the gate down a grassy slope on the left is a small aviary. Nearby, Clarence, the septuagenarian Galapagos tortoise, with his impossibly long neck stretched full-out, is likely to slowly turn his rocklike head, as if it moved on hydraulics, to look at you. On the right is an outdoor theater with a bare earth stage and metal bleachers. A short walk farther, there are a few low, unassuming buildings. This is the hub of the zoo, where the front office and the two classrooms, Zoo 1 and Zoo 2, are located.

The zoo stretches the length of the ridge, which makes for stunning views of the surrounding mountains and foothills and near constant, welcome breezes. An oblong-shaped road circles the teaching zoo's animals. Down the western side, the road is made of blacktop, lined with benches, and shaded by willowy pepper trees. This is the front road, where the public is allowed to wander on Saturdays and Sundays. The road turns to gravel along the zoo's eastern flank. This is the back road. The entrances to the animal cages can be found here. Consequently, this is the road favored by students, and you hear them crunching down it all day long.

Although EATM opens the gate to the public on the weekends and for field trip after field trip of kids lugging backpacks, it is first and foremost a school, which means it looks different from most zoos. In an age of lush zoos with enclosures that resemble African grasslands or polar ice floes, EATM has a bare-bones, old-fashioned look. Though the grounds are well landscaped with bird-of-paradise, cactus, roses, trumpet vines, mulberry trees, and oleander bushes, the animals are in cages, and there are no efforts to conceal that. The Bengal tiger, Taj, glares at you from behind bars. There are no chapter-long labels explaining mating and eating habits or lecturing the public on endangered species or rain forest decimation. At most, there's the species name, and often not even that.

The cages may look as if someone just emptied their office trash can in them. Student keepers are forever stocking the enclosures with all sorts of things to keep the animals busy and stimulated. This is called behavioral enrichment or B.E. In the parrot cages, there are hefty Los Angeles Yellow Pages with sunflower seeds tucked inside and rolls of unraveled toilet paper. Leafy branches are stuck through the primate cages for the animals to pluck, and big blocks of frozen Tang are left on the floor for them to lick. There are bowling balls for the cougars. Goblin, a baboon, cuddles a menagerie of stuffed animals.

You'll see other things you'd never see in a regular zoo. You might spot a student with a French manicure reaching her fingers through the bars and digging her shiny nails into the head of a hyena that's gone limp from the joy of a good scratch. You might see a gibbon, its long slender arms stretched through the cage, meticulously picking bits of dried skin off a student's arms. You might see Legend, the wolf, out for a stroll on her leash or Harrison, the Harris hawk, winging back and forth between two trainers. The teaching zoo is a lab where students can lay their hands on many different types of feathers, scales, and fur, and that is what makes the program so unusual.

So far during orientation week, the students have gotten to touch only two animals: the easygoing Chilean rose-haired tarantula, Rochelle, or any of the three-inch-long Madagascar hissing cockroaches. More people want to hold the tarantula that bites than the cockroaches that don't. These are two of the very few animals the first years will be able to touch until well into next semester. In fact, they are not even allowed to speak to most of the zoo's residents. If Taj chuffs at them, they may chuff back, but only one breathy greeting. They can talk to and handle the rats, the bunnies, the sugar gliders, the chinchillas, the opossums, all the reptiles, and Nova, the great horned owl. That's it for now. Sometime next winter they can say hello to Zulu, the mandrill, and Benny, the ancient capuchin. Until then lips must remain zipped. Students may not even linger in front of a cage long enough for an animal to notice them or, in some cases, make eye contact. The first years are like the cleaning staff of a high-end hotel, busy but invisible, and any interaction with the well-turned-out guests is absolutely forbidden. They clean the animals' poop out of their cages, but they mustn't utter a word. If a parrot says "Hi," the first years must turn a deaf ear, no matter how rude they feel.

There are a number of practical reasons for this legendary rule. The animals are generally unnerved by the arrival of new students each August. For example, Rosie, a baboon, is likely to scream and shake her cage like a berserk mental patient, the first years are told, until she gets used to them. Too much attention from these new faces may not only upset the animals but also distract them from their training regimen and cause them to lose their learned commands. They may confuse "Circle!" with "Speak!" or start waving when they should sit. The rule is especially important for the primates, who, as social creatures go, can be mighty prickly. For them, eye contact can translate as aggression. Look at a primate too long and he'll think you want a piece of him. The pissed-off monkey may then vent his rage on his unwitting trainer. Moreover, the rule is a test that the first years will do as they are told and that they will exercise great discipline in the face of great temptation. Everyone is rightly anxious about this rule. For these students, talking to an animal is a reflex as natural as breathing. The first years worry that without thinking they'll utter a "Hi, cutie!" or a "Hi, sweetie!" and the next thing they know, they'll have their walking papers.

"Don't talk to anything," Chris Jenkins, a second year, tells two first years, while they are standing in front of George, the fennec fox. George looks like a fairy-tale character curled in an afghan, softly whimpering for attention. George not only provokes an overwhelming urge to say hello but also to cradle him in your arms like a baby and sing a lullaby. "It's very unnatural to not say 'Hi' to animals, but everyone is listening," he continues. "There are severe consequences. It affects your grades, your animal assignments, it could even get you kicked out. Every year there's a couple who do it."

It is now Thursday, the fourth day of orientation week, and Jenkins, one of the best liked and most respected students in his class, is giving two first years another tour of the teaching zoo. This one is a behind-the-scenes look. The two first years, Susan Patch and Linda Castaneda, have had their pens and notebooks out all along the way. Jenkins starts at the aviary at the front of the zoo, where he tells them the turacos "like to buzz heads." The noisy guy is the plover, and the pheasant doesn't have a name—"just pheasant." Jenkins points out that Clarence, next door, is the only animal first years are allowed to hose off.

Jenkins leads them across the front road and behind the outdoor theater, where Happy, the alligator submerged out of sight in his pool, and two tortoises live. Here, Little Joe, much the smaller of the two, mounts Tremor and thrusts away, his hooked mouth opening and closing. Both Castaneda and Patch briefly quit taking notes. "It's like turtle porn," Jenkins says before turning to go backstage of the school's outdoor theater where he slides open the lid to a box, and several sets of eyes blink in the light. He's awakened the sugar gliders. Next, Jenkins points out the rabbits in their cages. If it's over 100°, put a water bottle in with them, he says. They stroll past the front office and through the squawky ruckus of Parrot Garden, where, Jenkins warns, you can accidentally get locked in the cages. As the threesome pauses by the emu's enclosure, Jenkins explains that when they clean her cage one student goes in and holds the emu down. Both Castaneda and Patch quietly contemplate this, while looking at the towering bird with her monster-sized feet. "It sounds worse than it is," Jenkins says.

Now they are in what's called the Show area near George. It's so hot that nearby Buttercup, the only trained badger in the country, Jenkins tells them, lies flat on her back, paws in the air. "She looks like roadkill right now," he says. "She's dug a hole nine feet deep." A trainer in the cage with Hudson keeps chanting, "Good beaver." They step up to the pen where Hamilton, the Yucatan mini-pig, lives and look down on his tattered black ears. He's not so mini, weighing in at 180 pounds. There's a rock in his pool, Jenkins says, because he likes to push it over. Castaneda and Patch scribble down the detail.

Next door, they duck into Carnivores, the area that everyone calls Big Carns, where three cougars, a Bengal tiger, an African lioness, a coyote, a wolf, a hyena, and two servals sleep most of the day away. The servals both bare their fangs and hiss. "Watch your fingers with them. They'll get you," Jenkins says. "The locks with red paint mean only staff can open those cages." They continue down the back road, and then through another gate, where the smell of a farm, that earthy mix of poop and hay, and sea lion barks fill the air. This area is called Hoofstock, which the sheep, deer, a miniature horse, the pigs, camels, and Schmoo call home. Jenkins demonstrates how the two camels can lean way out of their enclosures and trap you in the far end of Hoofstock. "These guys you need to watch out for."

Jenkins looks at his watch. He is due to walk Olive, the baboon, so he rushes through the rest, quickly ducking into the Reptile Room, a converted railroad car that smells like overripe fruit. He tells them to make sure the tarantula has water in her bowl, that none of the reptiles are venomous, and don't let Morty, the Burmese python, get too hungry. If so, he'll think his student handlers are dinner and squeeze them too hard. They sprint through Nutrition, a building devoted to making the animals' meals. He shows them the terrarium, where the mealworms are grown for the reptiles' dinners. They speed walk around Primate Gardens while the capuchins squabble and chatter. "These guys are hair grabbers and pullers," Jenkins says. He stops at the back of Primate Gardens and points to one last gate, this one between the lemurs' and the binturongs' cages. That gate leads to Quarantine, where a mishmash of animals—a raven, a snapping turtle, an arctic fox, a dog, among others—live. Then Jenkins is off, as Castaneda and Patch still write, hoping against hope to commit at least a small chunk of this avalanche of information to memory.

Patch is a lean, poker-faced beauty. She has a matter-of-fact manner and an athlete's grace. Blue rhinestone studs run up both her earlobes. Castaneda is nearly six feet tall and striking. Like many tall women, she slouches slightly. She has very pale, lightly freckled skin, black hair, and cultivates a kind of hip, bookish look. She wears a CamelbaK, the mouthpiece of which is draped over her shoulder and leaking water down her T-shirt. Both women are fairly typical of EATM students these days. They are in their twenties, they already have bachelor's degrees, and they are using this two-year associate's program as a kind of graduate program.

Patch, twenty-two, grew up in San Diego, where she regularly visited the city's world-class zoo and early on set her sights on working there. Her young life has been one calculated move after another to that end. Despite low pay, zookeepers' jobs are highly competitive, especially at a zoo with the reputation of San Diego's.

Patch began building her resume at the University of California, Davis, where she studied animal science, cared for the school's barnyard of animals, and trained a cow to be led on a halter. To beef up her animal experience, Patch volunteered at the San Diego Zoo, giving health exams to boas and iguanas at the reptile center. She worked part-time for her mother, a research scientist, plucking ovaries out of mice lying prone under a microscope.

By comparison, Castaneda, twenty-seven, discovered her affinity for animals relatively late. She grew up in Lynwood, where the wildlife was limited to rats and pigeons. She didn't see a squirrel or a woodpecker, she says, until she went to Lewis and Clark College in Portland, Oregon, on a scholarship. She started out premed, "so my parents could say they have a doctor in the family," but decided that animals needed more help.

After college, Castaneda moved back to Los Angeles with a degree in biology and taught science to grade schoolers and high school freshmen on her home turf, the inner city. She signed on for a summer gig as a research assistant helping a professor in Cameroon study the habits of hornbills and primates in the rain forest. She had to walk the eighteen miles into the jungle camp and ran out of water with four miles to go. The days in the rain forest were long, the nights uncomfortable and buggy, but the trip convinced Castaneda that her future truly belonged with animals. When she got home she applied to EATM.

Castaneda essentially moves through the world incognito. She's a Mexican American who grew up in the inner city, but people assume she's middle class and white. This leads to some awkward situations. People make racist remarks to her—outright ones or backhanded ones like "You could totally pass for being white." Even other Hispanics think she's white and doesn't speak Spanish. Her father is a retired machine operator who never really learned to speak English. Her mother, on the other hand, not only learned to speak English, but earned a high school degree, then a bachelor's, and finally a master's in linguistics. Nearly all the girls Castaneda went to school with, she says, got pregnant. Castaneda's family and work ethic made her different. "When my mother was my age she had three children and didn't speak English," she says. "I have no children, no house payment. I have no excuse for failure." So, like many second generation immigrants, Castaneda is under some pressure to succeed, even if it's self-imposed.

Both Patch and Castaneda are smart, confident young women. Neither worries that she will wash out of EATM. Both are aiming for straight A's. Patch thinks her UC Davis degree will stand her in good stead at EATM. Castaneda knows she can work harder than most anyone. Besides, after her trip to Cameroon and teaching, Castaneda feels especially unflappable. "What can EATM do to me that a fifth grade punk already hasn't done?"

Tonight yet another social function is on the schedule, the last one of the busy week—a bonfire. As tradition has it, the second years are the hosts, so they pack up a load of firewood, boxes of graham crackers, cans of chocolate icing, and marshmallows and head west to the broad beaches of Oxnard, where the temperature is a good thirty degrees cooler—a welcome change from the day's heat. After a good-sized fire is stoked, the students settle down in a kind of circle, second years on one side, first years on the other. I settle on a neutral spot in the circle where I have first years to one side, second years to the other. The conversation is, naturally, about animals. A second year tells me that she dreamed of being a dolphin trainer when she grew up in landlocked Wyoming. Another says that when her parents offered her the choice between the traditional Mexican coming out party or a horse for her fifteenth birthday, she chose a horse. A first year from Georgia tells me about volunteering at a shelter for fawns.

A woman had cleared all the furniture out of her house and filled it with cribs for orphaned fawns. When they jumped out, the first year would put them back in their cribs. "I totally love deer," she gushes.

If the first years feel a little as if they have intruded on someone else's party, they are not to blame. Many of the second years talk among themselves and ignore the first years. They crack inside jokes and sit close to each other. By comparison, the first years sit stiff and quiet. They don't know the second years or each other. They look over the flames at their very confident, very comfortable upperclassmen. As they've already been told on several occasions by staffers, this class of second years is exceptional and may be the best class ever.

It could be the second years are feeling too giddy just now to be good hosts. For them, the worst of the program is nearly behind them and they are about to come into the full wealth of being second years: field trips, later mornings, much more time with their animals. Essentially, they will be promoted to middle management. They will oversee all the scrubbing and make sure the first years follow the rules. This is why many of the EATM grads do not have fond memories of their second years. They are their bosses. This class of second years has sworn among themselves that they will be nice to their first years, mentors rather than shrews. But I've been told every class of second years makes this solemn pledge to themselves. All these students are here to learn animal behavior, but it's human behavior, often their own, that is most likely to trip them up.

As the evening deepens and the surf grows louder, the students go around the circle giving their names and ages. Then a game of telephone is begun, quickly becomes off-color, and then peters out. Marshmallows are spiked on foraged sticks and dunked into the fire, where they flame like mini-meteors, shooting sparks of burnt sugar into the black night. The flames reflect in young eyes that are eagerly, nervously turned toward the future. As the fire's crackling punctuates the lulling song of the surf, some of the first years wonder if they will shine; others, if they will make it. All of them consider the possibility that in the next year, month, or even week, they may be kicked, bitten, or scratched.

By Friday morning, the first years have dropped several hundred dollars at the bookstore on weighty textbooks on animal anatomy and an array of sweatshirts, tank tops, and polo shirts with the EATM logo. Now it's work day, the last day of orientation week and the traditional scrubbing of the teaching zoo from one end to the other. Both classes of students, numbering ninety-seven strong, will march through the compound with mops and buckets in hand. As the thermometer again hovers around 100°, students take brooms and knock cobwebs off evergreen bushes. They sponge the black mildew off the walls in the Reptile Room and pull a small tree out of Happy's sun-baked enclosure. All twenty-six bins of rodents are placed on desks in Zoo 1 so the Rat Room can be hosed. Legend, the wolf, stands and watches as three blond students wrestle a huge stainless steel sink down the front road.

I find Terri Fidone, a first year with dark eyes and a long black ponytail, at the back of the zoo, shovel in hand, contemplating an unenviable task. She has been assigned to scrub out the Dumpsters. The problem is they are full, and the truck that could empty them is off on an errand. If the truck doesn't show up soon, Fidone will have to climb into the Dumpster and shovel out the trash. "If I have to, I will, but I will be really bummed out," she says in her low voice. Until recently, Fidone was working on her bachelor's in biology while serving cocktails at the world's biggest casino, the mammoth MGM Mirage in Vegas. She started waiting tables when she was seventeen, then graduated to cocktail waitress. The tips were on average $100 to $200 a night—more if she worked the baccarat room, less if she worked the nickel slots. Once, a high roller at the roulette table threw her a $1,000 chip. When she dropped an organic chemistry class at the University of Nevada in Las Vegas because she was failing, she made a pact with herself to use the time she would have been in that class productively. She noticed a billboard for Keepers of the Wild, a sanctuary for exotic animals about an hour's drive southwest from Vegas across the Arizona state line, and signed up to volunteer there. While cleaning out the big cats' cages, she discovered that she "could shovel crap for eight hours and I'd rather be there than go to work and earn money."

After nearly ten years, she quit her job at the MGM in July to attend EATM. Now she's contemplating life as an EATM student, which means no more $100 sushi splurges or buying brand-name groceries. It will be generics for the next two years. "I have to get used to going to the bank to get money," she says. "Before, I always had cash on hand." A voice on the intercom asks if anyone knows where a pickax is. The truck arrives and Fidone is spared. She goes looking for sponges and a bucket.

The cleaning frenzy continues around me. Oddly enough, there's a lot of fussing over insects. The Reptile Room managers call out, "Ooh, spiders!" as they wipe down the walls. A threesome in the iguana enclosure hops and chants, "Yuck! Yuck!" as an ant colony they've disrupted streams up their bare legs. By the front office, Castaneda, still wearing her CamelbaK, points out a tarantula wasp to me, the big furry spider's only predator. It's about as big as a praying mantis and has a shiny black body and alarmingly orange wings. As Castaneda calmly explains to me how the wasp can paralyze a tarantula, the thing flaps its menacing wings, lifts off from the lawn like a helicopter, turns in our direction, and buzzes past us. Castaneda holds her ground and says calmly, "There it goes." Other students nearby—women who aspire to pet tigers and handle boas—shriek, duck, and run for it. Someone yells, "What the hell is that?" Turns out animal people are not necessarily bug people.

As the sun grows stronger, people zigzag like drunks as they walk from patch of shade to patch of shade, toting buckets and mops. They stand in the parrot misters. They squirt each other with hoses. Anita Wischhusen has sweated through her baggy tank top, a large wet stain spreading across her back, another pooling between her breasts. Despite the heat and hard work, she wisecracks loudly and cackles huskily at her own jokes. At forty-five, she is the second oldest student in the class of 2005.

Wischhusen is naturally loud but thoughtful. She has a gray, fuzzy mullet haircut and a dirty-brown tan. Her car is covered with pro-animal bumper stickers, even one that proclaims her membership in People for the Ethical Treatment of Animals (PETA), which is a kind of dirty word around EATM, but Wischhusen is not one to fit in. She grew up wanting to be a vet but didn't have the heart to euthanize animals. Her parents pushed her toward a secure career, so she went into computers. Her career may have been secure, but her lifestyle was not. When she wasn't at the keyboard, she was partying as hard as a Hell's Angel. At forty, after another night of drugging and drinking, she nodded off at the wheel, plowed into a highway median, broke her jaw, and landed in a court-ordered thirty-day rehab program. "The program told me to stop drinking, but did not tell me why I drank," she says.

When her brother came out, it occurred to Wischhusen for the first time that perhaps she was gay too. She'd always dated men and had sex "and all that, but it didn't do much for me." Maybe, she wondered, that was why she drained whole bottles of amber tequila at a sitting. In short order, she was not only sober but also had a serious girlfriend, a traffic cop with a daughter. At last, her personal life was in order. Then California's robust tech economy wilted, and Wischhusen was laid off from three jobs in a row.

After the final pink slip, Wischhusen thought back to her younger self, the one who dreamed of working with animals. She decided to go to EATM. That she is here is a small miracle. It took a couple of applications to get in. In the meantime, her father died of melanoma, the treatment for which used up the money her parents said she could have for EATM. Her girlfriend's daughter suffered a brain aneurysm. Wischhusen, who was teaching the girl to speak again, worried that she now had neither the money nor the time for EATM. Her girlfriend pushed her to go. Wischhusen sits for a moment in the shade, dabs at her wet shirt with a towel, and says happily, "I have no life now."

The first year students aren't the only new faces at the zoo this week. Amber Cavett, an outgoing, athletic-looking second year, leads me to the far end of the zoo through the gate at the back of Primate Gardens and into Quarantine. Here, we find Samburu. Samburu is a caracal, a small, fawn-colored African cat with tassels of black fur that droop decoratively from the tip of each ear. An EATM faculty member drove him from Northern California through the cool night and deposited him here this morning. The eleven-year-old cat is in the cage next to Tango, an arctic fox. While Tango pants loudly, the caracal lounges on a bed of hay in a box mounted about five feet up one wall of his cage. Word is that he's ornery, even nasty, but he seems quite demure just now. This is Cavett's first look at her new charge. She will train him for a grade this semester. She's fascinated but apprehensive because of his bad reputation. She's hoping to teach him to go into a crate on command—nothing too fancy. "They have round eyes rather than slitted like other cats because they hunt birds," Cavett says, looking at him. "They aren't nocturnal. They are crepuscular." You wouldn't know it by her upbeat nature, but Cavett has had one tough summer. Chance, the huge binturong (a native of Southeast

Asia that resembles a kind of raccoon on steroids with a huge prehensile tail to boot), bit her finger while she was showing off some training for her mother. He sank his teeth into her middle finger down to the bone. She couldn't bend it for two weeks. In June, she became the emu's lead keeper. Every day since, Julietta has tried to attack Cavett. When the 200-pound bird feels peevish, which is all summer, she rises up, stretches out her strange, turquoise neck, and then whacks at the nearest human with her beak. As if that wasn't bad enough, Julietta will kick with her three-toed feet. Whenever Julietta came after Cavett, she jumped back, even yelped. Cavett was told she had to get over her fears or she couldn't be Julietta's lead keeper. So Cavett drew on her experience as a scholarship runner at a community college in Oregon, where she put on a brave face for races even when she was terrified of losing. Cavett suppressed the urge to run for her life when Julietta attacked, stepped near the emu, and grabbed her blue, hairy neck. "If you stand really close to her, she can't kick you," she says.

We head to Nutrition to fetch Samburu his first meal at EATM. There we find various second years busily making dinner for their charges. They pluck whole rabbits from the walk-in freezer. They scoop crickets out of a terrarium. They unscrew jars of bright orange and green baby food. Cavett takes a cleaver and whacks off the head and feet of a frozen yellow chick. She pulls the stringy fat from three chicken necks, then chops them into bite-size chunks. She measures out 40 grams of canned horsemeat. Back in Quarantine, Cavett steps between Tango's and Samburu's cages. She will feed him by hand to develop a bond with him. She squats and tentatively holds out a piece of chicken neck in her fingertips, reaching through the bars at shoulder height. With a loud hiss, Samburu explodes out of his raised box, bounds to the bars, bares his teeth, and snatches the chicken neck from Cavett. "Okay, I'm a little scared," she says, holding her ground but turning her face away from him.

Samburu zips back into his box, jumps the five feet down to the floor of his cage, and bounces back up, hissing and growling loudly. Cavett holds another piece of meat through the bars. The cat leaps back down, grabs the meat with his teeth, inhales it, runs in a quick circle, and then suddenly, surprisingly, lies down on his stomach. He places his front paws neatly under him and gobbles up everything Cavett holds out to him. He gets up every once in a while and runs around the cage, only to settle right back down for more chow. "See that circle behavior?" Cavett asks. "I could put that on cue." Behind Cavett, Tango, perhaps inspired by Samburu's ruckus, gives himself a nice scratch. He runs back and forth along a series of wooden scrub brushes wired to his cage. The yellow bristles pull loose strands of his plush white fur. Tango pants happily. Samburu keeps hissing, but now it's a light, steady, contented-sounding drone. Cavett smiles. The summer is nearly over. The first feeding couldn't be going any better. This cranky, crepuscular cat with the fancy ears might just change her luck.

Behaviors

Animal training has been around since the first caveman threw a chunk of grilled mastodon to a hungry wolf lurking at the shadowy edge of the campfire light. Since then, we have taught dogs to herd, falcons to hunt on command, elephants to log, cheetahs to walk on leashes, lions to pull chariots, chimps to roller skate, and parrots to sing "Yankee Doodle Dandy." During nearly all that long history, no one truly understood training. Whether a trainer used brute force or kindness, nobody knew exactly why a technique worked or didn't. Animal training was like alchemy. That was until Ivan Petrovich Pavlov of the drooling dog and B. F. Skinner of the Ping-Pong-playing pigeons came along; they found the science behind the magic.

While researching digestion, Pavlov noticed that his research dogs drooled at the sight of white lab coats. That got the Russian physiologist thinking about involuntary reflexes. Could they be induced? he wondered. To answer that question, he placed a harnessed dog in a soundproof room. Each time a bell sounded, a puff of food powder was blown in its mouth. The dog would salivate.

Eventually, the pooch's brain linked the sound of the bell with a lip-smacking whiff of the powder. Then, all it took was one ding to make the dog slobber.

On the surface this seems a useless bit of information, but Pavlov's discovery had profound meaning for the fledgling field of behaviorism. He proved a reflex could be provoked through association. He dubbed this learning to connect one thing with another a conditioned response or reflex. It goes on all around us; for instance, a patient winces at the sight of a syringe or a kitty prances to the kitchen at the sound of a can opener. The behavioral equation is simple: A = B.

Pavlov's theory opened the door to a room that Skinner furnished. In his 1938 book, *The Behavior of Organisms,* Skinner introduced his new theory, *operant conditioning.* Through countless experiments with pigeons at Harvard University, Skinner demonstrated that behavior is shaped by its consequences. If a pigeon was rewarded for something it did, the bird would repeat that behavior. If the cooing bird gets a bit of grain for pecking a piano key, you can bet the bird will peck that key again and again and again. As long as the grain is forthcoming, the winged virtuoso will play on. What animal trainers had long assumed, Skinner demonstrated was a scientific truth.

Skinner's research also proved that rewards worked better than punishment. Not that punishment didn't change behavior; it did but not always as desired. A pigeon, or a person for that matter, is likely to look for ways to simply avoid the punishment. For example, police punish drivers who go above the speed limit. That discourages a number of lead foots, but a good number continue speeding with an eye out for police cars. Skinner found that rewards got the job done more effectively. For a species wedded to punishment, this was, and still is, revolutionary.

Skinner elaborated on the principles of operant conditioning in detail and came up with a whole vocabulary as well as equations and graphs with elegant curves. Along the way, he taught pigeons to play "Take Me Out to the Ball Game" and to walk in figure eights. He even taught the birds how to guide a missile by pecking at a target. He hoped his findings would make for a kinder world for his own species. Though that goal eluded Skinner, he changed animal training for good and for the better. He proved scientifically that a chunk of meat could generally accomplish more than the crack of a whip. Operant conditioning also gave animal training a level of seriousness, a scientific heft, it had never had. He took animal training out of the realm of the carny, the outsider, and plopped it down in the lab, where it might finally get some respect.

The wild animal tamers of yesteryear, a chair in one hand and a whip in the other, are now just images in old circus posters. Today we have exotic animal trainers. No enlightened trainer worth his clicker speaks of *tricks*, a word that conjures up the politically incorrect big top and the darker aspects of animal training. Animals are taught *behaviors*. They aren't kept in cages but *enclosures*. Times have changed and this linguistic shift reflects that.

Over the past two decades or so, the number of trainers using the principles of operant conditioning and positive reinforcement has grown steadily. Julie Scardina of SeaWorld says it's still not the dominant technique, but it is more and more broadly used. Clicker training, which is based on operant conditioning, has taken the dog world by storm and is on the rise among horse and bird trainers. Ken Ramirez, who oversees the training staff of the Shedd Aquarium, says, "When I look at where training was when I started in 1976, and where it is today, it's quite an evolution." Moreover, the whole idea of how to use training has changed. Training is no longer limited to the circus or the movie set. It is used in wildlife sanctuaries, zoos, animal shelters, research facilities, and educational programs. In his weighty manual, *Animal Training: Successful Animal Management Through Positive Reinforcement,* Ramirez writes that training today is "the cornerstone of a good animal care program." Scardina says that through training "you can make an animal's life safer, more positive and more interesting."

Training is now used to teach animals to cooperate with medical procedures, such as blood draws and ultrasounds—procedures for which they once would have been sedated. They are trained for general care; for instance, an elephant is taught to hold its foot up for a needed filing and dolphins are taught to give blowhole samples. Training is used for mental stimulation and exercise. Captive animals are trained to walk on leashes so they can stretch their legs and get a change of scenery. Gary and Kari Johnson start every morning at their private elephant compound in Southern California by leading their herd of pachyderms, each trunk holding the tail of the animal in front of it, on a bracing walk. Animals are trained so they can travel or go in front of crowds without undue stress. David Jackson of Zoo To You, a well-known educational outreach program in California, is often asked what drug he gives Jasmine, his Bengal tiger, that she'll lie still at his feet during his presentations. "People don't realize that she's been trained to do that," he says. "She's so calm and comfortable, nobody has a clue."

Enlightened trainers learn the animal's natural history to better understand the species. It makes a difference that a lion is a social animal and a tiger is not, that a horse is a prey animal and a wolf is not. Trainers study up on a species' natural behaviors so they know what will come easily to an animal and what won't. For example, baboons jump in the wild as a way to make themselves look bigger. In the open ocean, wild dolphins arc high out of the water just as they are trained to do in shows. Trainers learn an animal's physiology so they know what it can and can't do. Elephants cannot jump, though they can stand on their back feet.

As Ramirez writes, a contemporary trainer is "a keeper, a biologist, a mammalogist, an ethologist and much more."

Added to that is the fact that training is no longer thought of as a means to dominate an animal but rather as a means to communicate. Through operant conditioning a common language can be found, a mostly nonverbal one. As Karen Pryor says, this is a two-way system: "It is really an eerie thrill when the animal turns the training system around and uses it to communicate with you." Al Kordowski, an EATM grad and longtime marine mammal trainer with SeaWorld, explains it this way: "Imagine your best friend couldn't talk but you find a way to communicate and both achieve something together. That to me is it," he says. "We have a hard enough time communicating with each other. You've achieved something most people can't achieve in their day-to-day life, and you've done it with a wild animal."

EATM is credited with producing many of the foot soldiers in this training revolution. The program has graduated legions of trainers versed in operant conditioning and progressive animal care. These grads have infiltrated the animal industry from top to bottom, changing it as they went. Netta Banks, an EATM grad who works for the American Humane Association, monitoring animals on film and television sets, says, "Ninety-seven percent of young trainers are from the Moorpark program. They have raised the bar for zookeepers, for animal trainers and wildlife educators." The program's grads used to go primarily to marine parks, studio companies, and the circus. Now, more and more EATM grads land jobs at zoos and aquariums, thus spreading their training knowledge into those arenas. The school has also taken a formerly secretive skill and made it an academic curriculum at a public college open to all. In the past, aspiring trainers broke into the profession by cleaning cages at the circus or for a movie trainer until luck gave them a chance. There was no direct career path. Once EATM opened its doors, that changed.

* * *

Gary Wilson has taught the behavior and training classes at the school since 1985. Though he's nearly fifty, Wilson is still handsome in a boyish, clean-cut way. He runs regularly. He has a low-key but friendly, talkative manner. He seems more the thinker in a world of doers—someone as drawn to abstract ideas as he is to animals, who's as happy to discuss evolutionary biology as how to get Nick, the miniature horse, to blow a party horn. He teaches Sunday school at the Methodist church that his wife and two children attend. He's prone to giggling and has a wry sense of humor that can be lost on his students. He often wears a belt buckle with a lion's head emblazoned on it.

Wilson is an alum of the program, as is his wife, Cindy, who also teaches at EATM. He graduated in 1977 with Scardina. Both applied for a killer whale training job at SeaWorld. Scardina got the job and became the second female trainer at the park. Wilson got a job at Six Flags Magic Mountain, an amusement park, where he walked a lion on a leash. The lion, which had been in Tarzan movies, was missing the tuft of fur at the end of his tail. Wilson would lead him through a petting zoo, then clip him on to a front porch, where he slept the day away. People could pay $5 to have their picture taken with him. "People would hand you their baby to set by the lion, and you'd ask if they wanted to get in the picture and they didn't want to get near the lion," Wilson says.

Wilson then worked for two years with the Navy training dolphins. One of his cetacean protégés, K-Dog, was deployed in Iraq in 2003. He didn't find the Navy or training dolphins intellectually stimulating enough and went back to school. Eventually, he earned a master's in biological sciences from the University of California, Santa Barbara, in 1985. He started teaching at EATM while he was still working on his bachelor's, then joined the faculty full time after earning his master's. "My interest in training animals was an interest in helping animals," he says. "I saw this job as a chance to have a greater impact."

Like most marine mammal trainers, Wilson strongly emphasizes using positive reinforcement. He's an optimist whose leitmotif is "Try it." As one student tells me, "What's so awesome about him is that he thinks anything is possible." He's famous among his students for thinking up pie-in-the-sky behaviors for them to train. It's rumored that he once suggested that a student teach one of the school's capuchins to ride in a small basket hung from a helium balloon. When I ask Wilson if he did, he says he doesn't recall that, but it sounds like a fun idea. A few of the other trainers on the EATM staff think he may be overly optimistic and not cautious enough with the animals. He has, in fact, been attacked a couple of times. In one class graduation program he was described as the "animal chew toy."

Wilson first broaches operant conditioning in Animal Behavior, which the students take during either of their first two semesters. Wilson says the principles of operant conditioning aren't that hard to learn, but applying them is a different matter. That's why in the lab that accompanies the class, students train rats to run a maze. They learn to use a clicker, a small novelty store noisemaker, to tell an animal when it does the correct behavior, as a dolphin trainer uses a whistle. Wilson also lectures on the honeybee's waggle dance, the songbird's sub song, and the eel's long migration across the Atlantic Ocean. He speaks of filial imprinting, altruism, and the pecking order in chickens. He depicts the endless, intricate behaviors that, like the inner workings of a clock, make the animal kingdom tick.

In the summer, Wilson teaches a class in training using Pryor's *Don't Shoot the Dog!* and Ramirez's book. For that class's lab, students train two of the zoo residents for a grade. As with the rats, they nearly all work with treats, rewarding the animals when they do something right. Students meet one-on-one with Wilson on a regular basis to discuss their progress or any problems they are having. In the following two semesters, when they are second years, the students work with different animals. By graduation, the students have trained a hoofstock, a primate, a carnivore, and a bird as well as cared for a dozen or more animals.

Wilson finds that some people are natural trainers. They have an eye for body language. They are patient and calm. Trainers often refer to this as "animal sense." Plenty of trainers think that you have to be born with it and that, like common sense, it cannot be taught. Dr. Peddie is in this camp. Wilson disagrees. He finds some students are not natural trainers, but learn the skill from coursework and practice. Sometimes this latter group can outshine the natural trainers, who just try to get by on their instincts. The students who polish their inborn talent with what they learn in the classroom make the best trainers, he says. Then there are some students who just never get it, no matter how passionate they are about animals or how hard they study. "When it doesn't come together for some, I joke about dental hygiene school," he says.

Training is generally harder than most people assume, especially people who consider themselves good with animals. As one trainer put it, "It tweaks your head psychologically." It's like a Zen Buddhist lesson in self-control. You must be an intense observer, yet always think ahead, anticipate the animal's response, what one trainer refers to as "proactive second guessing." You have to think on your feet while always remaining calm. You must be confident yet cautious. If you make a mistake, you could teach the wrong behavior or even get hurt. If you get it right, as Kordowski says, "It's an adrenaline rush."

For the fall semester, Wilson has the students train husbandry behaviors, meaning anything that would help with the animal's care. One student has decided to teach Zulu, the mandrill, to offer his arm for an injection. Another has been assigned to teach Gee Whiz, the llama who doesn't like to be touched, to stand still and be brushed. One has set her sights on lifting Spitz, a serval, in her arms. Mary VanHollebeke has chosen, to Wilson's thinking, one of the hardest training assignments: to get near Starsky, the teaching zoo's Patagonian cavy. That may sound like a small order compared to teaching a baboon to be jabbed willingly with a hypodermic needle, but there are few animals at the zoo as skittish as the cavy.

The cavy spends his days on alert, twitching his nose ever so slightly with his ears pricked. Everything scares him. On the arid grasslands of southern Argentina he could flee a predator, sprinting up to fifty miles per hour, but in a cage—even a roomy one—there is nowhere to run. The large South American rodent looks like a rabbit crossed with a small deer. He has bunnylike eyes and large, upright ears. His slender legs are long and the toes on his feet resemble hooves. He has a fawn-colored coat with white along his belly and between his back legs; his rump is edged with a slight skirt of fur. Starsky arrived this past spring with another cavy that soon became sick and died. That left Starsky, an animal that usually lives in pairs or threesomes in the wild, alone—that is, except for VanHollebeke.

Starsky is new to the training game, as they say at EATM. VanHollebeke is the first student to work with him, and that alone should make her task slow going. Animals eventually understand training in concept and that if they do as asked, a reward is in the offing. An animal brand new to training, such as Starsky, is typically confused by the whole process.

VanHollebeke began this past summer, sitting very quietly by a tree about ten feet away from his cage, letting the cavy get used to her. In training talk, this is what they call desensitizing. She held perfectly still, which doesn't come naturally to VanHollebeke, who charges through the day until she collapses into bed at night. If VanHollebeke moved, even to push a loose hair away from her face, the big-eyed rodent would dash to the far end of his cage. If he got really scared, he'd jump madly against his cage, even scrape himself. And so VanHollebeke took her sweet time.

Once the cavy got used to VanHollebeke in one spot, she'd move a step closer to his cage. If the cavy was unbothered, she would reward him by taking a step back. Slowly, VanHollebeke worked her way toward the cavy, inching closer and closer. When she got within a few feet of his cage, she taped a spoon to a long stick. If VanHollebeke stepped closer and Starsky held his ground, she would slip the spoon on a stick through the bars and deposit a snack of alfalfa pellets on the ground.

By the end of the summer, the second year was finally able to slip inside the cavy's cage and sit on the concrete ledge near the door without scaring Starsky too badly, though he kept his distance. Now, as often as she can, VanHollebeke sits in his cage. Perched on the ledge, quiet and still, VanHollebeke says she does a lot of thinking. Meanwhile, the cavy hides behind a bush at the far end of his cage and occasionally peeks out with one large, unblinking eye.

VanHollebeke is a tall, dynamic beauty with an oval face and straight dark hair that reaches far down her back. She is an unlikely candidate for a training task worthy of a monk. She likes things to be perfect and to go as planned, and when they don't, she has a short temper and a sharp tongue. She's much more calm with animals, she says, and is drawn to the distressed, difficult ones. VanHollebeke, like many EATM students, originally thought she'd be a vet. She was putting herself through college in Michigan, going part time and taking pre-vet courses. One night her mother couldn't sleep, turned on the television, and saw a program about EATM. Though it was 4:30 A.M., she woke her daughter.

A year into the program, VanHollebeke has found that training is not for her. She does not, she admits, have the patience for it, especially all the repetition. She likes caring for the animals, especially making their lives more interesting with behavioral enrichment. She brings Starsky corn on the cob and browse. She stacks up his hay in a pile so he can busy himself stamping it back out.

VanHollebeke also likes a challenge, even one within herself. What better way could there be to learn patience than by ever so slowly convincing a South American rodent to trust her? Seated in the shade of Starsky's cage as he tucks his small, fleet figure behind the bush, VanHollebeke seeks a serenity that has eluded her. She also dreams the impossible—of touching this Patagonian cavy before she graduates next May.

September

For the two weeks following orientation the new students take notes until their fingers cramp. It's the worst injury they risk just now. Each morning, they follow Brenda Woodhouse around for a crash course in running the teaching zoo. Woodhouse, who's trained dolphins at Chicago's Brookfield Zoo and worked at the bird show at the Los Angeles Zoo, has a job similar to a drill sergeant. She must get her new recruits up to speed fast. She has to instill in them a long list of the dos and don'ts of zookeeping at EATM. She has to teach them how a zoo is in the details, that the smallest oversight—one forgotten parrot feeding, one tool absentmindedly left near a primate's cage—can have huge, even tragic, consequences.

A new keeper hired at a typical zoo would have to learn how to run only a portion of it. At EATM the first years have to learn an entire zoo as fast as they can. When cleaning the chinchilla's cage, they should let him take a dust bath for a minute or two. Zulu, the mandrill, gets a cup of juice in a *paper* cup. Three times a week the algae on Sally, the snapping turtle's shell should be gently scrubbed off.

The first years learn the three *L*s: life, liquid, locks. A keeper should always check to make sure an animal is alive in its cage, that it has water, and that the cage is locked. Woodhouse shows them where the red emergency phones are. They learn why they must always wear a whistle. If an animal gets loose, you blow your lungs out. If you hear a whistle, duck into a classroom, a storage shed, or even an animal cage. Woodhouse tells them if a primate grabs them through a cage, not to scream or yank, which is fun for the monkey. If a big cat attacks you, roll into a ball. If a camel knocks you down, roll away from it. If there's an earthquake, all the students who can should report to the teaching zoo. If there's a fire, there's no emergency plan for that just yet. It's coming soon, Woodhouse says.

Woodhouse shows them how to put on the miniature horse's halter, which is simple enough, how to walk the emu, which can require some running backward, and how to tie a falconry knot with one hand. They learn how to pick up the big snakes so their body weight is properly supported. If they grab just the head and the tail, they are told, the snake's spine may snap. It's a lot to take in all at once. Woodhouse knows it. She teaches with a big smile, lots of encouragement, and a voice that naturally booms. She's got a maternal touch, but no one doubts she means business, especially after she begins marking students tardy. The first years are due at 6:30 A.M. sharp. It's an adjustment. They speed through the early morning fog, blitz into the parking lot, barely turn off the ignition before they have a foot out the car door, then sprint for the front gate, where they fumble with the latch, losing precious seconds. When the classroom clock's hands reach 6:30 A.M. exactly, Woodhouse starts reading down the roll.

On the first Wednesday, Larissa Comb, a Colgate University graduate and a former Wall Street stockbroker, pulls into the parking lot with only minutes to spare. She's about to dash for the front gate when she notices a second year taking her time. Comb must have more time than she thinks. She slows her pace. What she doesn't know is that Woodhouse reads the names alphabetically. Comb also doesn't know that the second year strolling through the parking lot's last name begins with *Z*.

When Comb slips into the classroom, Woodhouse has just read through the Cs. Comb grabs a seat and doesn't say anything until Woodhouse has read the entire roll. Though she was only thirty seconds late, Comb is docked for the amount of time lost until she announced her presence—three minutes. She should have yelled "Here!" as soon as she stepped in the door. The longer the tardiness, the more docked points. Comb just lost two. That's a pittance, really. What worries Comb is that just ten days into the program she can already kiss her sterling record good-bye. Her heart sinks. This is why: In the winter the first years will ask for the animals they would like to work with come summer. The staff considers the requests based on the students' grades and attendance. Straight-A students with perfect attendance get first dibs. The students with less stellar records get the leftovers. Comb just slid down a rung.

When not following Woodhouse around the zoo, the first years hole up in the air-conditioned chill of Zoo 2, the larger of the program's two classrooms. In Anatomy and Physiology (A&P), Dr. Peddie, like an old fire-and-brimstone preacher, strikes the fear of zoonotic diseases into the first years' hearts. With relish the vet regales the first years with the symptoms of rabies, bubonic plague, and his personal favorite, Valley Fever. "I tell them you can end up wearing diapers," he chortles to me. "That is always an attention grabber." His point is simple—wash your hands, wash your hands, wash your hands.

The average first year takes seven classes. In Animal Diversity they begin to work their way through the animal kingdom slide by slide, starting with sponges (phylum Porifera), jotting down the family, genus, and common name of each species as they go. In Animal Behavior they dive into the theoretical underpinnings of operant conditioning and how to care for a rat. Gary Wilson teaches both classes. He is a notoriously tough test giver, who will mark off an answer if it isn't worded exactly so. Second years have warned the first years not to take more than one "Gary class" per semester. In addition to the class on zookeeping, Woodhouse runs a class on conservation and another on wildlife education.

Classes have started for the second years as well, though not nearly as many. In Dr. Peddie's vet procedure class they study skin, from calluses to carcinomas. In Wilson's training class they discuss how, through bad timing, you can accidentally teach an animal an unintended or unwanted behavior, what is called superstitious behavior. The second years still spend most of their time running the teaching zoo as they did this summer.

From August to early May, both classes, 100 or so strong, work at the zoo, but once the second years graduate, half as many students are left for the same job. This summer, as with most summers here, the mercury regularly shot over 100°. A few students fainted from dehydration. They stood in the walk-in freezer in Nutrition to cool down. Under the sun, the animals' poop bloomed to a pungent ripeness, especially in Hoofstock, where the camels and sheep scatter droplets all day long. When they weren't cleaning, feeding, or training, the students were putting on education shows for zoo visitors.

The summer has exhausted them. The second years try not to lick their collective chops or count the hours until the first years take over the lion's share of hosing, raking, poop scraping, and chopping up dead chicks. However, this also means sharing what has gotten to feel like their zoo. The latter may prove harder than the former.

* * *

On the last day of August, a Sunday, the baton is finally passed. The first years fan out across the zoo and go to work. Linda Castaneda spends her first week cleaning in Hoofstock. When Castaneda goes in with the sheep, they give her a once-over, bumping her with their soft black noses, inquisitively baaing at her. She refrains from baaing back. She starts to smell like "butt." There's a mystery brown spot on her shirt. When her boyfriend does their laundry, an animal turd rolls out of the cuff of her pants.

Being brand new to zookeeping, the first years immediately begin to make mistakes. Castaneda leaves the wheelbarrow too close to a camel's enclosure. One first year forgets she shouldn't let the guenons get overhead while she's cleaning their cage. Ramon, the male, pees on her head. While scrubbing the Reptile Room, a first year plops Dot, the yellow leopard gecko, on her shoulder, thinking she'll stick there. She doesn't. The gecko tumbles to the floor. Dot survives that free fall only to have another first year on the very next day plop her on her shoulder; again Dot tumbles tail over snout. The first years, humiliated, learn their lesson: not all geckos stick.

Michlyn Hines, who's in charge of running the teaching zoo, comes across a first year who has hung a lock on a cage where Samantha, the gibbon, can reach it. Hines no sooner finishes telling the girl about the lock than the first year begins sweeping about a foot from Samantha's cage, where the gibbon, a notorious arm grabber and hair yanker, could easily reach her. Already there's animal interaction—too much of it. A first year is repeatedly caught chatting to the lemurs. One first year insists on leaning her face in close to the mandrill's cage, a no-no. First years greet Abbey, the dog, when she's out for her morning stroll. They are all told to zip it. It's largely up to the second years to enforce the rules and that power can't help going to some people's heads. Some, even though they said they wouldn't, get a little bossy and a little superior. They scold when cages aren't clean enough, when logs aren't filled out correctly, or when the tortoises' overhead light isn't turned on. They make the first years raise their arms to see if their shirts stay tucked in, per the dress code. They report the first years' mistakes to the staff.

This makes the first years alternately skittish and resentful. The first years complain that second years talk to them as if they are stupid. A good number of the students have worked with animals before and don't appreciate being treated as if they know nothing. One chafes when a second year tells her how to roll up a hose. Anita Wischhusen, forty-something, with a full résumé of professional jobs, is written up by a second year half her age and half her size for not "taking direction." Wischhusen shoots her dirty looks whenever she can and bad-mouths her. First years can't help noticing the second years breaking a rule or two, answering their cell phones on the back road, or going into animal cages without the required second person to back them up.

At student council meetings, Becki Brunelli, a second year, tries to keep the peace. She reminds her classmates to be understanding with the first years. At thirty-three, she's one of the oldest in her class. She has an easygoing manner, but nothing gets past her. She's tall, has a head of long, kinky brown hair and a lilting voice. She loves rats and has one tattooed on her ankle. She is a strict vegetarian. She is one of Dr. Peddie's favorites. He's cooking up a way to hire her at EATM after she graduates this spring. That would suit Brunelli just fine.

Maybe because she's so capable and because she could do most anything she set her mind on, Brunelli has had trouble settling on a career. She was an elementary school teacher. Then she went back to school to study anthropology. Brunelli did that for two years and then switched to computer science. She got a job as a Web designer for Warner Brothers. She made a sizable paycheck there but found that so many long days in front of the computer wore her down. She got out of shape. Her natural ebullience waned.

On a whim, Brunelli went on a trip to the rain forest. Not long after touching down in Iquitos, Peru, she felt younger and healthier as she soaked up the oxygen-laden, moist air. Each night she fell into a sleep so deep that the din of the nighttime rain forest never disturbed her. She realized she had to escape the cage she had made for herself, her fluorescent-lit cubicle, and make a different life. When she returned to Los Angeles, she resolved to change careers again. This time it would be animal trainer.

At EATM, she's gotten nearly straight A's, except for one B in behavior lab class, where her rat would not run its maze. She's a class leader. She's as happy as anyone who's happened upon their calling. However, a few things do nag.

As Dr. Peddie warns all students, EATM often trumps love. Over the years, the program's demands have caused its share of breakups and divorces. When students' grades slip, one of the first questions Dr. Peddie asks is "Do you have a boyfriend?" If the answer is yes, "Are you having trouble with him?" Though Brunelli's grades don't show it, she is. For the first year of the program she lived with her boyfriend, an actor, but they saw so little of each other, she decided this summer to move closer to the school and do away with her hour-long commute. They haven't broken up. In fact, Brunelli, always the optimist, hopes the change will improve their relationship.

Brunelli's other distraction is the number of credit card bills that land in her mailbox each month. She came to EATM with a goodly amount of debt, chiefly student loans. Then it took her some time to adjust from a $70,000 salary to the near zero income of a student, and her bills have only gotten bigger. This debt weighs on her, especially when she thinks of her newly chosen profession. Entry-level pay is low for many animal jobs, about $8 an hour. Having found her calling, Brunelli wonders if she can afford it.

For now, she throws herself into life at the school. She is on three committees, including the student council, does all the zoo's scheduling, and cooks up schemes to raise money for the school. This semester she's assigned to Rowdy, the skunk; Nuez, the agouti; Malaika, the African gray parrot; and Goblin, the thirty-year-old hamadryas baboon. Goblin is her favorite, though Brunelli hates to say so. The baboon has a sweet disposition and spends most of her day grooming her stuffed animals. She loves yogurt so much she will throw it up into her mouth for a second, third, and fourth taste. Brunelli often sets up a chair next to Goblin's cage so the baboon can trace every follicle, mole, and bump on her forearm. It's so relaxing that the baboon goes into a meditative trance and Brunelli falls asleep. Brunelli is training Goblin to let her insert an ear thermometer. She's already taught the baboon to hold her ear near her cage bars. Now, once a day or so, Brunelli crouches down and pokes her finger into the baboon's ear, getting her used to that first.

After a long day at the teaching zoo, Brunelli often goes home at night to find Harry, an orangutan, sitting in her living room. Her roommate works at the Universal Studios animal show and brings the primate home for sleepovers. "I never get tired of animals," she says. "No matter what's going on in the rest of my life, I can go hang out with animals. It's always therapeutic." In other words, it's hard to worry about your love life crumbling or impossibly high debts when you are watching TV with an orangutan.

The teaching zoo's inhabitants are the constant at EATM. They come and go, but many of the inhabitants have called the teaching zoo home for most of their lives. Laramie, the one-winged golden eagle, arrived in 1977. Koko, the forty-one-year-old capuchin, moved in in 1975. Precious, the anaconda, has lived here for almost thirty years. It's the students that change regularly, and this is not lost on the animals. Even though the temperatures are still hot enough to nap most of the day, the teaching zoo's residents notice the new, inexperienced keepers. Taj, the Bengal tiger, eyes flashing green, back slung low, stalks the first years as they clean by her cage. In Primate Gardens, Samantha lets loose with her warning call, a slow building whoop. Rosie, the olive baboon, screams at any new student who dares to walk near her cage.

Most of the teaching zoo's animals are donated. Some are gifts or loans from movie trainers, such as Abbey and the camels, Kaleb and Sirocco. Some come from private owners who found that a Bengal tiger or a serval do not make a good pet. Some are abandoned or rescued animals that need a home, such as Buttercup, the badger who was found at a Moorpark gas station when she was a baby. She had a puncture wound. There are as many hard-luck stories as you'd find in a Dickensian orphanage: Happy, the alligator, is about half the size he should be, because a man raised him in a bathtub and fed him fast-food hamburgers; Olive, the olive baboon, was illegally captured and confiscated while still a baby; Tyson, the military macaw, has breathing problems from being stuffed within a car door to be smuggled into the United States.

On the flip side, some of the EATM animals have impressive resumes. Puppy, the turkey vulture, was in *Airplane*, sitting on Peter Graves's shoulder as the plane dived. Schmoo was in *The Golden Seal* when she was only six months old. For years, Schmoo was the only trained sea lion in the business. She appeared in the TV series *Dharma and Greg* and did commercials for Frigidaire, DuPont, and Saran Wrap. Sirocco, the white camel, appeared in the movie *The Mummy*.

As in most zoos, many of the animals at EATM are in their golden years. This keeps Dr. Peddie as busy as a doctor in a rest home. Louie, the ancient prairie dog, has lost the use of his back legs and pulls himself around. Koko shakes from old age. Schmoo is nearly blind from cataracts. Her hearing is going. She is epileptic, for which she takes phenobarbital. She also has ulcers. She takes Zantac for that. In all, she gets twelve pills and vitamins every day, all of which are stuffed into the fish she eats.

Dr. Peddie's job is made easier and harder by the students, who are quick to tug on the vet's sleeve. "If an animal hiccups twice in the wrong way I'm alerted," he says. He has to filter out what are honest concerns and what are overreactions. Like new parents, the students have a tendency to worry. "We have all the rejects, the ancient ones, and the students keep falling in love with them," he says. As usual, several of the zoo's inhabitants demand Dr. Peddie's attention. The rabbits have walking dandruff. Their bedraggled coats are filled with Cheyletiella mites. Samuel, the corn snake, has five eggs stuck in her uterus. When you turn her upside down, you can see the lumps. The first surprise was that Samuel, who everyone assumed was male and thus the name, laid a clutch of a dozen or so eggs last spring. The Reptile Room managers suspected she had more in her, but the snake ate and defecated through the summer, so they didn't say anything to the vet. Recently, Samuel became irritable, so the managers spoke up. Dr. Peddie gives the corn snake a medication to make her pass the eggs, but none of them budges. Then the vet tries to draw them out with a syringe, but the eggs are, as he says, "too cheesy." There is one last option: squeezing them out with his two big hands. Dr. Peddie puts that task on his long to-do list.

Most worrisome is Zeus's case. Zeus is a female green iguana, who, like Samuel, was believed to be male until she started laying eggs. No one knows her age but the assumption is that the iguana's old. Certainly Zeus moves likes she is, mostly dragging herself around by her front legs, her back legs limp behind her. The reason becomes obvious when Dr. Peddie holds up a set of X-rays in class. The vertebrae in her lower back have fused together, possibly from a bad break. Dr. Peddie shows the students the X-ray, not just to teach them but also to prepare them for what comes next—putting her down. The vet is in no rush. When it comes to euthanizing an animal, Dr. Peddie believes in the art of diplomacy and consensus building. He will hold off putting an animal down, if it's not in too much pain, until the majority of students are ready. Sometimes the staff, even students, think he's too slow to pull the trigger.

By the last week of the month, the first years sag under the weight of classes and morning cleanings. Dr. Peddie confides to me that he's a little worried, that this new class is "fragile." The first years are on average younger than the second years. They appear to have more personal problems. The second years complain that they aren't at the teaching zoo enough. Woodhouse thinks they are doing okay, but admits the first years are not jelling as a class. They are getting a rep, like every class of students that has passed through EATM, but it is not a good one. This pains the overachievers, like Castaneda. But who could possibly measure up to the second years? The staff adores them. For the lowly first years, resentment builds. While the first years slave away like Cinderellas, the second years have all the fun. They dissect rats; they cavort with the animals; they run out for big bags of fast food; they go to scuba class. It also doesn't help that the second years don't invite the first years to their parties and that they sometimes don't even acknowledge them when they walk into a room.

The first years, Castaneda tells me, "feel like they are drowning." Many first years are surprised by the program's academic rigors. For Wilson's Diversity test, the students need to memorize forty-four slides with each animal's family, genus, species, order, and common names in that exact order. Everyone carries flashcards all the time, glancing through them as they rush down the zoo's front road. A detailed drawing of the entire zoo is due in Woodhouse's class. A test in Dr. Peddie's Anatomy is in the offing.

Students who'd given up their ambitions to be vets because they didn't fare well in science classes find themselves again struggling with science classes. Even a student like Comb, who got good grades at Colgate without killing herself, says, "This is the first thing I've had to work hard at. It's a different kind of difficult. In college I used to take a break and play my violin. Here I'm too tired. Every piece of you is exhausted."

One first year teeters on the edge, having come late to the morning cleaning several times. The first years look to their left and right and wonder who won't make it through the first semester. Add to that, the Newcastle quarantine has ended, and that means there are pigeons to be had. Any day now, the first years will be expected to break some bird's neck with their bare hands.

Nutrition

Humans have a deep urge to feed animals. Note the bird feeders that dot our yards and the piles of oily french fries left for noisy gulls. We leave out food for foxes, deer, opossums, rabbits, and squirrels. Though it's illegal, more and more people feed wild dolphins in Florida, Texas, and the Carolinas. Some people even pay for the chance to serve sharks dinner. National parks warn visitors not to give leftovers to bears, mule deer, moose, coyotes, prairie dogs, marmots, and so on. This human drive is not limited to our borders. Down under, the Tasmanian Wildlife Service warns visitors not to feed wallabies, currawongs, and Tasmanian devils. At zoos, people are forever chucking candy in with monkeys or tossing marshmallows over the fence to polar bears, despite signs asking them not to. Eric Baratay and Elisabeth Hardouin-Fugier write in *Zoo: A History of Zoological Gardens in the West* that on a day in 1957, visitors to Antwerp's zoo fed an elephant 1,706 peanuts, 1,089 pieces of bread, 1,330 sweets, 811 biscuits, 198 orange segments, 17 apples, 7 ice creams, and 1 hamburger.

Why all the feeding? Food is a common currency. Every living creature needs it, craves it. A handout is the surest means to some kind of interaction with a wild animal. Moreover, food is love for humans. For us, offering a morsel is an engraved invitation to cross-species friendship, though an animal sees the exchange in a much more utilitarian way. This need to feed is deeply satisfied at EATM, where the students are supposed to feed the animals.

All the food prep goes on in Nutrition, a building that seems to constantly echo with the thwack of a dull knife and the kerplunk of monkey chow dropping into stainless steel bowls. As a radio blares, EATM students hurriedly pulverize melon, scramble eggs, and rend heads of romaine. The room is shiny with stainless steel counters. There are shelves packed with jars of brightly colored baby food, plastic bins of freckled bananas, box after box of oatmeal. Nutrition is perfumed with the thick aroma of slightly overripe fruit. Much of the produce is donated by area grocery stores, and so the lettuce may be slightly wilted, the grapes a bit shriveled, the yellow squash dimpled with rot.

Though the room is filled with food, you can quickly lose your appetite here. In the large walk-in freezer there are frosty, headless pigeon bodies, plastic cups of pink baby rats, and buckets of thick, soupy blood left over from thawing horsemeat for the big cats. The students whack slippery chicken necks into chunks. They collect mealworms and crickets. They squeeze the remaining yolk out of the pearly baby chicks with their fingers.

Prep work here can be straightforward or a creative outlet for the students. Food provides the building blocks for a lot of behavioral enrichment. Students artistically lay lettuce leaves and strawberries on a tray for the tortoises, which prefer brightly colored produce. Students stuff meat inside coconut shells for Savuti, the hyena. They freeze blocks of blood for the big cats to lick with their rasplike tongues. Even a slight change in preparation makes for B.E., so the students cook rice in coconut milk or dice vegetables extra small. They offer whole pieces of fruit or a snail still in its shell. All these culinary variations break the routine for the students as well. Behavioral enrichment is a two-way street.

For most animals, the students need only to prepare the correct amount of food, deliver it to the cage, and keep the records up to date. For a few others, feeding is a little more involved. When serving the snakes their very occasional rat, the students have to remember not to handle a prey animal beforehand. Snakes hunt with the nose, and if your hand smells like a rabbit, they may bite that. If you are assigned to Schmoo, you will be forever thawing, rinsing, and sorting fish, not to mention removing small squid from the pen. If you are assigned to either of the two eagles, you'll be setting traps for squirrels and rabbits around the teaching zoo and then gassing them to feed to Laramie or Ghost. Then there is Nick, the not-so-miniature horse.

Nick is, as an EATM staffer puts it, "oinky fat." A year and a half ago he weighed 285 pounds, according to zoo records. Over a six-month stretch the mini-horse's weight soared to 337 pounds. Now he hovers around the 300 mark. That's still too hefty. His student trainers could cut back his food, but that's not the problem. The problem is that Nick's a thief and, worse, he's stealing from his best friend.

At the teaching zoo, Nick is a popular animal with the students and the public. He knows about sixteen behaviors, including sticking his tongue out and turning in a circle. He is good-looking, with a caramel-colored coat and a long blond mane that hangs dashingly across his forehead. Nick came to the teaching zoo in 1997. He arrived with the llama, Gee Whiz, or Whiz, as a lot of the students call him. The hooved twosome were rejects from a petting zoo of sorts. Neither liked strangers touching him. Whiz has quite an underbite. He has large eyes, and his coat resembles a big, rumpled sweater. Regardless, Whiz has a regal bearing. He stands much taller than Nick, though the horse is wider.

Hoofstock generally do better with company, so the twosome live in the same corral near the camels, the little pigs, and the mule deer. Nick and Whiz are boon companions. They often stand side by side. Nick, however, is not the best corral mate. He steals Whiz's food. The llama has bad teeth. Trainers must moisten his pellets. Still, Whiz eats slowly, chewing in exaggerated, clockwise rotations while he stares ahead. Before Whiz can finish his dinner, Nick polishes off his own meal of oat hay and then sticks his handsome muzzle in the llama's bin for second helpings.

The second years assigned to work with Nick this fall semester are determined to slim the tubby horse. They have started him on a new exercise regimen. In addition to his daily walks around the compound, they now take him into the teaching zoo's smaller amphitheater, remove his bridle, and let the boy run. Like a rodeo horse, Nick tears around in mad circles and figure eights, kicking up a spray of dust and wood chips with his back legs as he goes. The commotion usually makes C.J., the coyote, yip and Legend, the wolf, howl in their nearby cages.

One of the student trainers, Jena Anderson, a tall young woman from Minnesota with big blue eyes and a deep, melodic voice, has begun retraining the horse to pull a one-person cart for a grade this semester. She's started by reteaching Nick to run wide circles while on a lead, what is called lunging. She finds Nick is out of shape. When she directs him to keep running, Nick looks at her as if to say, "You gotta be kidding me." "He's a moocher and a slacker," Anderson says.

Anderson is low-key but determined and accomplished. She has a degree in fine arts from St. Olaf College, taught art history, directed a church choir, and was a director of a vet clinic. When she called the Minnesota Zoo asking about careers working with animals, they told her about EATM. She spent a year working at a religious camp in California to get residency so that EATM would fit her budget. She describes herself as a recovering Catholic and "practicing Christian. . . . I don't go around saying praise Jesus," she says. While she was working at the camp, nine ministers took her out for a drink and tried to convince Anderson that she should go to divinity school. "That's not what God is telling me to do," she says.

If you had to bet on her or Nick winning this battle of the bulge, you'd put a big pile of money down on Anderson. She's not only exercising him more but also doing her best to outsmart the ravenous horse. She has moved Whiz's food trough out of Nick's reach, tying it high on a bar in the corral. That seems to be working, but there are reports that Nick has taken to ramming Whiz in the neck while he is chewing, making the llama drop his half-chewed meal on the ground.

Amy Mohelnitzky has the opposite problem. The second year puzzles over how to get a certain animal to eat her breakfast and, like a worried mother, tries her wiles in the kitchen, cajoles, and generally frets. The object of her concern lives at the back of the teaching zoo, where unexpected barks can surprise a visitor. This is where Abbey, the teaching zoo's briard mix, lives.

Abbey is one of the few animals at the zoo who could give Schmoo a run for her money in the training department. She's on loan to the teaching zoo from a studio trainer and has a long list of commercial credits, including advertisements for PetSmart. Abbey is also what Dr. Peddie calls a "difficult keep," which makes her a good addition to a teaching zoo. Her long, glamorous coat, which is a smoky black down the back and cream-colored around the face, ruff, and down the legs, requires constant attention to keep it untangled and flowing. A few days without grooming and Abbey begins to look like a Rastafarian who fell in a mud puddle.

Like many a starlet, the pooch has a finicky appetite and always prefers attention to a treat. Consequently, her svelte figure often becomes a little too svelte. When you weave your fingers through her coat, you easily feel her ribs and her narrow waist. Though she'll usually eat in the evening, most mornings Abbey turns her soft black nose up at breakfast. When the briard mix doesn't eat, she always throws up a puddle of watery, yellow bile. Then the pup doesn't feel up to training. This has been her story since coming to the teaching zoo two years ago.

Abbey, a dog living at a zoo, can't help looking out of place amid the big cats, iguanas, baboons, and macaws. She lives at the far end of the zoo because her excited woofs scare the other animals, especially the three spider monkeys that madly whirl around their cage and puff their chests out as Abbey trots by. As part of the first years' long list of zookeeping duties, Abbey is walked on her leash each morning. Her coat flowing, her pink tongue poking out, Abbey sashays past the screaming capuchins, the big-eyed camels, and the snoozing mountain lions, barking and barking as she prances along. The only animal she notices is Legend. Any student walking Abbey down the front road knows she has to tighten her grip on the leash and flex her biceps as they near Legend, because the two canids carry on like mortal enemies. Legend lunges at her cage and growls a low, toothy threat while Abbey pulls her leash taut, straining with all her might to get at Legend.

A dog at a zoo—even a studio dog with a long list of commercial credits—is a bit of a second-class citizen. Everyone loves her, but she is not an exotic, and that is what the students came to train. Many have already had dogs, and there is a feeling that training one when you could be training a Bengal tiger is a missed opportunity. Never mind that there are far more jobs training dogs, whether for the movies, for sniffing work, or for therapy, than there are jobs training exotics. Also, dogs, compared to the rest of the teaching zoo's inhabitants, are not dangerous. This was not lost on Mohelnitzky. While some of the EATM students get a buzz off working with dangerous animals, Mohelnitzky does not. She admits to being afraid of some of the animals, though she's as brave as any of the students. On stage one time, Kaleb, the dromedary, kicked his legs out in a mad, threatening jig, thrashing his head to and fro, and tossed Mohelnitzky around like a rag doll while she clung to his reins. She fell on her knee so hard that she couldn't walk for two days. Abbey not only offers a break from steeling her nerves but also the dog actually craves her attention. "Most animals are like give me my food and leave me alone," Mohelnitzky says. "Coming out of the cage is the happiest thing for Abbey." Besides, Mohelnitzky is head over heels in love with Abbey, and that love salves a wound.

Mohelnitzky is pretty, like many of the EATM students, though she's of average height; most of the EATM girls are tall. Mohelnitzky always looks put together. She wears makeup. She complains she's in the worst shape ever but doesn't look it. She's from Wisconsin. When she was nineteen, Mohelnitzky moved with her then boyfriend now husband to California. He's an actor who dreams of having his own talk show. Mohelnitzky came to the West without a dream per se. She had always loved animals but never thought there was any kind of career working with them. She studied nutrition for one year in college and got a job as a personal trainer in Santa Monica. She found out about EATM during a visit to the San Diego Wild Animal Park, quit her job, and began waiting tables and working in a vet's office so that she could take the prerequisites for the program.

During her first year at EATM, Mohelnitzky came home one night to see a blood stain on her front doormat. A workman at their apartment complex had let himself in to do some work. If they had known he was coming, they would have shut Barney, their little beagle mix, in the bedroom. They didn't. When the workman opened their front door, he let Barney out. The pooch ran into traffic and was killed. The couple was heartbroken. They moved.

Once again Mohelnitzky spends her days thinking about a dog. Abbey's previous student trainers didn't get her out of her cage much. She thinks that might be part of her stomach problems, that Abbey doesn't get enough exercise. She springs Abbey from the cage whenever she can and takes her to every class with her. When it's too hot or rainy, Mohelnitzky puts Abbey in the front office, even though some staff members grumble. She's started feeding her twice a day, rather than just in the evening, which had been the practice. Mohelnitzky wonders if she has an ulcer. These late summer mornings, she mixes cooked vegetables and a can of wet dog food into her kibble, garnishes it with a sprinkle of acidophilus powder, and heads to Abbey's cage to tempt this picky pooch's appetite. "If she were a tiger, I think more would be done for her," Mohelnitzky says, "but she's just the dog."

Animal People

Among the ninety-seven students currently enrolled at EATM there is a house husband, an eighteen-year-old vegan from Colorado, a plumber who used to earn $50 an hour, and a young New Yorker whose ultimate goal is to get a Ph.D. in evolutionary psychology. Most of the students are from California, but some hail from West Virginia, Iowa, North Carolina, Michigan, Washington, and even Peru and Argentina. A goodly number have bachelor's degrees. A few are in their forties and a handful are in their thirties. They are changing careers, getting that college degree that may have eluded them years back, working a midlife crisis out of their system, or all of the above. The bulk of the students are in their early to mid-twenties. A couple are fresh out of high school. Many of the EATM students are paying their own way, moonlighting as vet techs or manning registers in video stores, racking up credit card debt, and taking out personal loans.

Despite the differences in ages and backgrounds, the EATM students talk of feeling deeply at home, of finally being in their tribe. They are. They are all animal people. This may make the school a headier experience than it already is. Many of these students have always felt out of step with the world. Here, they can talk about animals all day and no one will give them a sideways glance. Being among their own has its downside, though: animal people typically aren't at their best with people, even other animal people.

The term *animal people* gets tossed around a lot at EATM and among professional trainers, but when you start asking what it means exactly, not everyone is sure. What becomes clear, though, is that the term is a double-edged sword—a badge of honor on the one hand, shorthand for misfit on the other. Gary Johnson sums it up when he tells me that an animal person is "someone who will work long and hard. It's whatever it takes to get the job done. Animals don't care about workers' comp or coffee breaks." Then he adds, "People who hang around animals tend to be on the strange side."

The simple definition of this term is that animal people are passionate about animals. Animals are not on the outskirts of life but at its very core. Typically, they have felt this way their entire life. One EATM student told me that when she was a kid and played house with her friends, she always volunteered to be the dog. Another got the idea to be a dolphin trainer when she was five, though her family had no pets other than a guinea pig and she grew up in a small town in Wyoming far from the ocean or a zoo. Others told stories of feeding the family pets and of nursing ailing bunnies.

Most children are strongly drawn to animals, but for some that feeling does not wane as they grow up but rather it intensifies. Any child watching a trainer in a shiny wet suit ride atop a killer whale might think, I'd like to do that, but an animal person holds tight to that dream. It's reflected in how students apply over and over to EATM (one as many as five times) or attempt to return after being kicked out. A number of professional trainers told me that that their job was not a career but an identity. "This isn't what I do," says Kris Romero, an EATM grad and staff member. "This is what I am."

Now it's a fine line between loving animals deeply and preferring their company to your own species. It's a line some animal people drift across, and this is where the misfit connotation comes from. Dr. Peddie puts it diplomatically, saying some animal people don't communicate well with people and so take refuge in the animal kingdom, telling their deepest secrets to their hamster or plastering their bedrooms with posters of horses and lions. Mark Forbes, an EATM alum who is second in charge at the studio company Birds & Animals Unlimited, puts it less diplomatically: "There are so many animal people that are horrid at dealing with people."

I found most animal people to be energetic, strong-willed extroverts, though I did run across some walking stereotypes. I found a surprising machismo for a profession that is dominated by women. Trainers rarely fail to mention how hard their job is, whether the physical work or the long hours. I heard the buzzword 24/7 over and over. "They take pride in how difficult it is," Dr. Peddie says. Trainers spoke of their profession as a calling on a par with joining a convent or monastery. Michlyn Hines, who worked at the Los Angeles Zoo for seventeen years before joining the staff at EATM, says in the eighties few women zookeepers had children; the job was considered too consuming to accommodate motherhood. I also found animal people generally opinionated and competitive. They can be quick to criticize one another for everything, from how someone keeps her compound to her training techniques. Karen Rosa, head of the American Humane Association's Film & TV Unit, hears studio trainers tear each other down on a regular basis.

As Brenda Woodhouse points out, animal training builds ego, and that ego gets the better of some trainers. Though they might be only a few months into the program, even EATM students don't shy away from offering scathing opinions of professional trainers, not to mention of each other. If I mention one to another, I often get the lowdown on the other student's failings: one has to have everything her way; another is a know-it-all; X is retarded; Y is a bitch; Z doesn't have a lick of animal sense. The second years get along famously and generally don't trash each other, with a few exceptions such as the classmate who has gotten a reputation as a pathological liar. When it comes to the first years, though, many of the second years let it rip: the first years this; the first years that. The first years often turn on themselves, grousing about their student council and gnashing their teeth over every misstep their classmates take. "There's more stupid people here than I thought existed," says Wischhusen of her class.

Like new mothers, EATM students are forever gauging who is the most devoted. There's nothing like nurturing to bring out women's competitive streak. Everyone notices when someone takes time off from the teaching zoo. "If she doesn't have to be at the zoo, then she's not here," is a recurrent criticism. If you are not there every day all day, you are considered a fair-weather animal person. Terri Fidone spends time with her new boyfriend on the weekend. Students notice. Wischhusen stays home with her girlfriend when she can. Students notice. There is tsk-tsking over a first year who sits in her car in the teaching zoo parking lot listening to the radio one afternoon.

That EATM students can be so fractious is ironic, because training and operant conditioning work on humans too. An EATM staffer tells me, "It helps you understand why people do the things they do. They are learning such good people skills. I'm sure they thought [the program] was all about animals." That lesson is, obviously, often lost on the students.

Does the school have a testy tone because there are so many animal people at EATM or because there are so many women?

A preponderance of X chromosomes usually makes for what one EATM staffer calls the drama du jour. There are cliques, plenty of hurt feelings, a steady stream of gossip, and roommate problems. The college's psychologist, Laura Forsythe, says at one time she was seeing three women, separately, from the same house. Forsythe says the psychologist before her said to expect a lot of EATM students. He was right. She finds the same problems—depression, eating disorders, even suicidal tendencies—that she finds in the overall student body but with more frequency among the EATM students. "People talk about it being a hothouse," she says. She hears similar complaints from students in the nursing program, which is also dominated by women. EATM students, Forsythe says, may suffer more because they are so isolated. The program is so consuming they have little time for a social life, let alone eating well. If only the vending machine had less candy and more high-protein snacks, Forsythe despairs.

For some of the students, though, the experience is a revelation. A number told me that they had always had trouble getting along with women until EATM. Here they find the best friends they've ever had. You can tell who's close to whom by where they sit in Zoo 2 during class. The buddies clump together in twosomes or groups of three and four. Loners, such as Wischhusen and Fidone, tend to sit in the very front row or toward the back.

That there are so many women at EATM comes as a surprise to many. The image most people have of animal trainers is men in fancy getups with big cats like Siegfried & Roy or the late Gunther Gebel-Williams of Ringling Bros. and Barnum & Bailey Circus. Also, women are not generally drawn to dangerous work, which animal training is, but then most dangerous work—mining, commercial fishing, truck driving—does not include caretaking.

Women have dominated the student body at EATM since the program opened its doors in 1974. There has been only one class with a nearly even gender split. In the school's early years, the female student body was a bellwether of change. By the eighties, more and more women became zookeepers, historically a man's job. They now represent 75 percent of keepers nationwide, according to a 2000 survey by the American Association of Zoo Keepers. In 1977, Julie Scardina was the second female trainer hired by SeaWorld. Now women count for more than half of the company's trainers.

The number of men has declined as the education requirements for keeping and training jobs has increased but the pay hasn't. Not so long ago, a zookeeper didn't even need a high school diploma, just the muscle to hoist a fifty-pound bag of feed. Now a college degree is expected for jobs that pay not much more than the minimum wage and still require scooping poop. This may have deterred men but not women, who are generally more willing to accept lower wages in exchange for following their dreams.

The preponderance of women at EATM makes for a strange hybrid—a boot camp crossed with a sorority. There are all the rules, the physical demands, the early morning reveille, the uniforms, and the class hierarchy of a military school. According to Dr. Peddie, the school is designed, like the Marines, to break the students down in the first semester, then build them back up again. However, a female faculty member burst out laughing when I told her he said this. Dr. Peddie may exaggerate, but the program is an ordeal physically and mentally. Mixed into the rigorous regimen are all kinds of girly rituals of parties, gift giving, and skits. "You're working with animal poop, getting sweaty, getting huge muscles, you have to do something that is girly," says first year Larissa Comb. Like Catholic school girls, the EATM students gussy up their uniforms with accessories. They strap on shimmering belts, pull on socks with paw prints, and clip on silver earrings in the shape of arcing dolphins. Fidone sports diamond earrings her new boyfriend gave her. Susan Patch has a small clock in the shape of a ladybug hanging from a belt loop. A first year wears striking woven belts she brought with her from Argentina. Some do go into deep schlump mode, hiding under caftan-sized jackets and pulling visors low over unwashed hair, but plenty of the students brush on mascara and even nail polish. They dab on lip gloss before taking the serval for a walk. They change their hairstyle or color as often as runway models. In these details they retain their femaleness, not to mention their individuality, as EATM grinds them down and builds them back up to the gold standard of animal people.

The irony is that before too long animal people may be on the endangered species list themselves. Working with animals is less and less a safe haven for misanthropes and introverts. Movie trainers have to work with a long list of people on the set. Most killer whale and dolphin trainers are performers. Even zookeepers are expected to set down their rakes and talk to the public more and more. A growing number of animal interaction programs, such as Discovery Cove in Florida, require trainers to work closely with the public. That is why at EATM students have to learn to speak in public and work as a team. At the school, they also learn how to behave within an overwhelmingly social hive. If EATM truly succeeds, by the time they graduate, these animal people will be people people.

October

On the first Saturday of the month, the EATM students leave their uniforms at home, slip into their civvies, and gather for a party in the San Fernando Valley at Trevor Jahangard's house. Jahangard is one of the two male second years and at twenty is one of the youngest students in his class. Some of the second years refer to him as "our baby" and tease him like a younger brother. He lives with his family, who regularly host huge get-togethers of their Jewish, Iranian, Mexican, and African American relatives. So a dinner party for some eighty EATM students and staff isn't that big a stretch. Long tables line the backyard. There's a ping-pong table in one of two living rooms and a pool table in the other. His mother has laid out a prodigious buffet of Persian food. There are platters of beef stew with lentils, cucumbers doused with yogurt, a chicken salad dotted with peas, and planks of pita. Pomegranate seeds shimmer like so many rubies in a large bowl.

What with lip gloss applied and hair curled, some of the students are almost unrecognizable from their EATM selves. Silver belly button rings twinkle and inky tattoos peek out from under low cut T-shirts. Some have brought boyfriends, so there are far more males in the mix, which seems to make for a mellower mood. Most of the students slouch on couches and watch various games unfold. Dr. Peddie picks up a pool stick. A number of students gather to heckle him. Linda Castaneda, I, and some first years hover near the buffet table. We snack on wedges of pita and talk of Siegfried & Roy. Just last night, Montecore, a seven-year-old white tiger, bit Roy Horn on the neck during a performance and dragged him offstage as if he were a felled deer. Horn's in critical condition. We debate the story the Siegfried & Roy camp has floated, that the tiger was somehow helping Horn. These novice trainers don't buy it. "He went right for his jugular. That means he thought of him as prey," Castaneda says.

Horn's attack is a visceral reminder to these young women that they have embarked on a dangerous profession. Even the most skilled trainers are at risk. VanHollebeke's mother has already called to tell her she can't work with big cats. Before she entered the program, Susan Patch's father made her promise that she'd never work with elephants. But you don't have to work with a powerful predator or gargantuan herbivore to put yourself in harm's way. As they say over and over at EATM, anything with a mouth bites.

Though the first years long for animal interaction, they mean the friendly kind, but you can't have that without tempting the toothy, snarly kind. There are hundreds of ways to get hurt at the teaching zoo. Brenda Woodhouse teaches the first years the finer points of keeping safe here and in their future jobs. She instructs the first years to locate an animal in its cage before going in, have escape routes in mind, and be ready to fight the animal if it comes to that. They should know the animal's behavior, Woodhouse says, so they can decipher its body language. An animal doing something out of the normal—say, roaming its cage when it normally sleeps—can be a bad sign. Watch out for a primate shaking its cage or bouncing up and down rigidly. Keep an eye on a parrot's beak and don't let the birds get above you. An angry tiger's eyes will flash green. A mountain lion is more aggressive than people assume. Always, always back out of a cage. Woodhouse can give pointer after pointer, but ultimately it's up to these students to find the midpoint between too much fear and not enough. Caution requires a careful balancing act that will come naturally to some and elude others, no matter how hard they try. The animals will notice if they are scared or careless. Some species will be forgiving; others will not.

Though the teaching zoo is run largely by people brand new to working with exotic animals, most injuries are inflicted by tools. There have been a few close calls. A student was bitten badly on the arms by a baboon. Another was mauled by a tapir. One student was attacked by a lion. This was in the 1980s when the teaching zoo was in cramped quarters down by the football field. As Gary Wilson tells it, the student was feeding animals in a row of cages, working her way toward the lions. The lions were in a large arena at the time. A tunnel led from their arena to their cages. It was blocked with a plywood board. In a fit of excitement, Hatari, the male lion, banged on the board so hard he broke the chain that connected the tunnel to the arena. The lion charged through the resulting narrow breach. The student with the food had her back to the big cat and never saw or heard him coming. The lion grabbed her from behind, sinking his canines into her neck and shoulder. A male student struck Hatari with a rake, breaking it over the cat's back, according to one alum. The lion let go and hightailed it to a corner of the compound. He hid behind a barn. Jamie LoVullo, who was then a student at the school, says, "It was the first time you learned that an animal you loved could kill you."

The injured student survived, but the bite caused nerve damage to her face. Whether EATM would survive wasn't certain at first. The student sued the school and the EATM staff, including the then director, Bill Brisby. They settled out of court. Wilson heard the student used the settlement to start a dolphin swim program in the Bahamas. EATM continued. Hatari was not put down.

All the students know the story of the lion bite, but over the years it has been turned around some, as many EATM tales are. There are rumors that she hadn't told the school that she was hard of hearing and a hemophiliac. Wilson says he doubts she was a bleeder; she resisted going to the emergency room after the attack, he says—not typical hemophiliac behavior. She was somewhat deaf, Wilson says, but as he remembers, it didn't make a difference. The students who saw the lion escape never had a chance to warn her.

Wilson himself has been attacked twice at the school. In 2000, Kissu, the declawed mountain lion, turned on him while he was in a caged area with the cat. It was orientation week. A film crew from *Animal Planet* was at the teaching zoo. Wilson was leading the cougar through a caged corridor to a play area. The cat dawdled. Wilson called, "Kitty, kitty!" Kissu turned and looked at him. His pupils had become huge, black discs. Wilson knew he was in trouble. He didn't have a bobstick, which was a mistake. Kissu lunged at Wilson and bit down on his elbow. He had on a sweatshirt and a jacket, so the mountain lion's teeth did not puncture him. Wilson knocked the cat off and grabbed him by the scruff. He tried to push Kissu into his cage, but the cougar charged back out and sank his canines into Wilson's rump. Wilson grabbed the cougar by the back of the neck again, and this time shoved him into his cage. *Animal Planet* caught the white-faced Wilson fresh from the attack as he told Dr. Peddie what had just happened. "What saved me was that he didn't have claws," Wilson says. A Dutch student raced Wilson in his BMW to the emergency room. "That was probably the scariest part of the whole thing."

Afterward, Wilson wondered what had triggered Kissu's attack. He had taken care of the cat in his home when Kissu was a cub and had spent hours sitting in his cage. He called Sled Reynolds, a professional trainer who lends the teaching zoo its camels. Wilson asked Reynolds why he thought Kissu had attacked him. He doesn't need an excuse, Reynolds told Wilson. "I tend to be more analytical than that, but he's probably right. Even though an animal is an animal you have worked with a long time, they can still turn on you."

Wilson still dreams of Kissu charging at him, his canines bared, his pupils like big black holes. It's not his only recurring nightmare. He also dreams of the time when Lulu, a camel, nearly killed him. That was more harrowing, he says, because there was nothing he could do. In the spring of 2002, Lulu delivered Kaleb at the teaching zoo. Wilson had stopped by the school with his wife, Cindy, and their two children to visit the baby camel. Some students and staff milled around. Before leaving, Wilson tried to milk Lulu. The camel, who Brenda Woodhouse says was grinding her teeth and seemed agitated, was in no mood to have her teats pulled. As Wilson leaned under her, Lulu kicked him with her front legs, knocking him to the ground. Then she sat down on him, pressing him into the hard-packed earth with her belly. Wilson heard his ribs crack. His arm was twisted. Woodhouse says all you could see of Wilson were his legs sticking out from under Lulu. "I was just pinned," Wilson says. "I thought this is a really shitty way to die. In the dirt, under a camel's belly. I can't breathe. I'll pass out."

People screamed and pushed. The camel wouldn't budge until a student stabbed her in the side with a Leatherman. Wilson escaped. His arm was hurt and some ribs were broken. The more serious damage may have been to his psyche. In recent years, Wilson has started to have panic attacks. He thinks it's due to years of sixty-to-eighty-hour-long weeks he pulled over the years, but the camel attack, not to mention the cougar attack, couldn't have helped. Still, having a camel try to kill you has its benefits. Always the optimist, Wilson says, "I feel fortunate I came close to death. I appreciate life more."

In actuality, it's the animals you least suspect that inflict the most wounds at the teaching zoo. This past summer, Sequoia, the demure-looking mule deer, beat up a second year who was training her. The student was in the deer's pen when Sequoia rose up on her back legs and clocked the second year with her front hooves. The blow cracked open the student's head and left dark bruises all over her chest and legs. Maybe the deer went ballistic because the second year had been in Big Carns and smelled like the cats. Maybe it was a jealous rage. The student was dating Sequoia's favorite human, the gardener at the zoo, who lovingly brings the deer browse and sits by her cage. "Deer aren't very nice," she tells me. "They can be very aggressive. Nobody told me that until I got on her."

Back in May, Birdman, the kinkajou, nailed Mary VanHollebeke. Kinkajous resemble fairy-tale animals, with big, luminous eyes, pertly pointed noses, and long curlicue tails. They are nocturnal, so when VanHollebeke stepped into Birdman's cage to leave his food, usually he snoozed away in his dark den box. That day she leaned over to set his bowl of food on the ground and stood up to find, to her surprise, the kinkajou out on his shelf. He bit her right hand and then her left, sinking his teeth deep into her flesh. He held on. VanHollebeke tucked Birdman between her legs and squeezed, so the kinkajou couldn't thrash. Another student tried to scare, poke, and pry Birdman off. He wouldn't budge. VanHollebeke could hear him slurping her blood. After a few minutes, Birdman finally let go. He'd bitten all the way through her hand. "There was a lot of blood," she told me. "It was all over." An infection took root in one hand. VanHollebeke had to go on an antibiotic drip every four hours. She wore bandages on both hands for two weeks. She needed two months of physical therapy to repair the muscle and nerve damage to her left hand. VanHollebeke ends her story by saying, "He's got the sweetest face in the world."

The most recent emergency room visit was the work of Gabby, the parrot. The macaws, with their sharp hooked beaks strong enough to break a finger, inflict many of the wounds at the teaching zoo. Students work with them at close range, and the parrots can be quite peevish. Gabby is the same parrot who punctured a student trainer's hand during orientation week. Two weeks ago, the parrot bit again, this time on the student's lip. The second year had leaned close to the bird for a kiss. Instead, she felt a sharp pinch, the warm ooze of blood, and a flap of her lip roll into her mouth. Her mother paid for a plastic surgeon to stitch her lower lip back together.

Students who have been munched wear their scars like purple hearts. They describe them with a mix of world-weariness and self-mockery. There is some relief in being hurt, in knowing exactly what it's like to be chomped or rammed, in knowing you survived, and in knowing that you are not infallible. An animal bite takes a bit of the ego out of these young trainers, which can be a good thing. It's the students who get bitten repeatedly that the staff worries about. They have their eye on one first year who has already been bitten a couple of times, including a chomp from a lemur.

The unscathed—nearly all the first years, including Castaneda, and some of the second years, including Brunelli—may dread the unknown. However, having skin that has never been breached by fangs or beak and never been bruised by horn or hoof may make you feel superior. Even the levelheaded Brunelli can't help feeling a tad proud of her no-bite status. To her, it's proof that she may be a natural with animals. This confidence is unlikely to survive EATM. Few students graduate without a nip, if not a puncture wound or two.

As for Horn, the general consensus near the buffet table is that the showman had been playing with fire for years by bunking with tigers, romping with tigers, and snuggling with tigers. They are, after all, tigers. Up until now, despite his talents, Horn had been lucky. And whatever happened on stage, Horn was very unlucky to have fallen down in front of his tiger. These trainers in training know enough to realize you never want to resemble prey to a large, toothy, muscular predator.

The conversation runs out of steam, so we head outside into the dark to see the mews Jahangard has built in a corner of his family's backyard. In addition to going to EATM, Jahangard is apprenticing to be a falconer, a consuming process on its own. The party is still going strong, though some of the revelers have to get up at 6 A.M. or so. On the back porch, a first year tells Dr. Peddie a fart story. He guffaws. Then he regales us with one of his classics, the one about Boobs Mackenzie. Not long after she'd had a botched breast reduction, Boobs brought in her dog. She said, "Dr. Peddie, would you look at this?" He thought she meant something about the dog. She lifted up her blouse. As he tells it, "She had her nipples pointing south, right at the floor." He laughs at his own story. As the clock nears 11 P.M., people start to say their goodbyes and drift toward their cars, as the prospect of hosing poop on too little sleep, and possibly with a hangover, begins to weigh on their young minds.

At 8 A.M. the next morning, a deep fog cloaks the zoo, softening the hard edges of the chain-link front gate. As I walk through the gauzy mist, a bespectacled female figure near the front office calls out to me, "You've come at a great time. We've got eighty pigeons."

That is eighty decrepit homing pigeons to be pulled. Eighty pigeons is an unprecedented number of birds; usually, the school gets a dozen or so at a time. The birds always arrive unannounced, but once here, they are quickly dispensed with. The owner of the pigeons, a guy who has been bringing them here since EATM started, lifts them out one by one and hands them to a second year who goes by the name of CVP. She's well liked by both classes. She has a round face and a very feminine voice that nearly trills. As the owner hands her the pigeons, he points out especially old ones for me to see, such as one with a very craggy waddle. I don't really want to see the birds, so I keep my eyes on my notebook. The pigeons coo and coo. CVP matter-of-factly loads them into two crates, then totes them down the front road to the hay barn, which is next to the Rat Room and across from Zulu's cage in Primate Gardens.

Word spreads quickly, and before long a small circle of apprehensive, quiet women in sweatshirts has formed by the barn. These nurturers have come to be executioners. CVP has set the two crates of softly cooing pigeons next to two large garbage cans. Just behind her in the barn, another second year has set up a butcher station. The gray light glints off the large cleaver with which she'll chop off the wings and feet.

"We're here for you," CVP says to the small circle. "If you want to cry, cry on our shoulders. Some people cry. Some people laugh. Try to respect how people react."

The small circle freezes in place. Some put their fingers in their mouths. CVP plucks a bird out of a crate and calms its fluttering wings. She wraps her left hand around its body. "It's dislocation. You'll pick up the bird and pull its head right off." With her right hand, she hooks her index and middle finger around the bird's head, so the knuckles of each hand are touching. "My hands are tiny," she says to herself. "Don't let them get loose. They're homing pigeons. They'll fly home."

She leans over and into the garbage can, pressing the bird against the side. "Blood will squirt out. Don't look down at the bottom of the bucket. They might blink. [The body] will twitch for about a minute and a half, so press it against the side of the can." And then unexpectedly, without missing a beat, she jerks her right arm and off pops the pigeon's head. We don't see it, but hear it thud on the bottom of the garbage can. "The quicker you do it, the better. The longer you take, the worse it gets," she says. "He's still twitching."

The small group frowns and shifts from foot to foot. There are a few doleful uh-uhs. Everyone is clearly uncomfortable, but there's not one tear or giggle. The first two volunteers step forward. One says, "I think I might laugh." They both lean over the can, birds in hand, follow CVP's orders, and quickly pull the heads off. They turn their heads so they can't see the bottom of the bucket. The one who thought she'd laugh smiles uncontrollably and titters slightly as they stand there forever, waiting for the bird's nervous system to still. When they hand the pigeon bodies over to be chopped, their hands are slick with blood. "Oh my god," someone in the circle says. Another two volunteers step up, including the bright-eyed first year from Georgia who took care of baby deer. "Give me an ugly one," she says.

As the cleaver chops away in the background, the two first years dutifully take the birds in hand, lean over the can, and pull. Then the Georgian lets loose with a kind of play by play. "Mine moves too much; uh, mine moves too much. He's trying to walk." She half smiles, half grimaces as she holds the jerky body against the can. "There's like juice all over my hand. It just shot up my pants." Waiting students stare at their feet and rub their hands on their pants.

Some more first years, including Castaneda, arrive just as this twosome hand over their pigeons and march off with their bloodied hands held in front of them like zombies. The cleaver thwack-thwacks in the barn. "Is that blood?" one of the students who just walked up asks. "I'm going to faint." Castaneda jumps in. "Can I get an ugly one?" she asks. Maybe because she missed

CVP's matter-of-fact introduction, Castaneda is the first student to look clearly upset as she takes a pigeon in her hand and assumes the execution position at the garbage can. So does her partner. CVP coaches them. When Castaneda pulls, the head doesn't come off; she has to jerk a second time. "Oh god," she moans. Her partner begins to cry. Castaneda wells up. They hand over their birds and wipe their eyes on their upper arms as blood drips off their fingers.

Just then, a slight, older student with bleached blond hair walks up. This is Chandra Cohn. She is already crying—tears running down her cheeks, eyes red, sniffling, the works. The mood quickly tumbles downhill. What was grim but businesslike suddenly becomes the scene of a tragic accident or massacre. Thwack! Thwack! the cleaver pounds. One pigeon sits in a crate by itself. It's so pretty no one will pull it.

"This is the hardest thing you'll ever have to do here," CVP says, trying to restore calm. She says this with the slick red hands of an ax murderer. When she hasn't been showing first years the drill, she's been pulling pigeons like a machine. There are just so many birds to kill, far more than the first years can handle. A couple of other second years have arrived to help and quickly begin popping heads off. "Whoa, got a squirter," one says. "Dude, this one is blinking."

The group of onlookers grows as a few staff members and second years arrive to watch. They mean well, but the onlookers add to the ghoulish spectacle. The student from Argentina with the stylish belts and earrings that dangle more than they should pulls halfheartedly with lips trembling, wailing, "I can't do it." A student standing back from the group near Primate Gardens lets out a deep sob and walks off. Another student pops a head off, then drops the bird's body, which thrashes in the can, its wings rustling the plastic garbage bag. She has to reach in and grab it. She opens her mouth so wide in disgust you can see her silver tongue stud. The pigeons in the crates grow quiet and still.

"Whoa, got a live one," calls the second year, chopping away as a headless body flaps across her cutting board. Somehow Cohn summons the nerve to step up to the garbage can, but when she pulls her pigeon she sobs uncontrollably and shakes. Next to her, CVP pops another head off. A very small first year with an upturned nose and a pretty, starlet way about her strolls up with her pants tucked into her wellies. She smiles and says, "I want to do it before I get sick and pass out."

"We know first aid," CVP says.

"I have a problem with anxiety," the first year says, smiling, and paces in a circle, while CVP collects a bird for her.

"You're going to realize you're going to do things you never thought you could," another second year volunteers.

The first year jumps from foot to foot. Finally, bird in hand, the demure student leans over the can—the smile, a bit sad now, still on her face. She turns her head, jerks her arm, and, poised, pulls the bird's head off.

Some of the students who went first have drifted back, their hands now clean. They've come for their pigeon bands. It's an EATM tradition to keep a shiny leg band from a pigeon you've pulled. Typically, people add them to their key chains. They are half badges of honor, half mementos of the sacrificed birds. The bands are also, well, pretty. "Can I have a blue band?" a student with eyes still red from crying asks. Even Cohn, who took it worse than anyone, returns for a band, though she's still sniffling. "The pigeon was the same color as my bird," she whimpers as she holds her hand out for the band.

The morning wears on and the hysteria passes. Eyes are dried. Notes are compared. Pigeon bands are wedged onto key chains. Castaneda tries not to think about the way she could feel the bird's vertebrae pull apart and how its warm body trembled in her palm. As I walk up the front road, I run into Anita Wischhusen. I ask why she didn't pull a pigeon. Because, she informs me, she has no intention of doing so—period. She's hoping for another outbreak of Newcastle disease. If not that, she'll find a loophole in the school's rules. Before I get to ask her why, we reach the front office and are within earshot of a staff member. Wischhusen goes quiet and walks away from me. She returns and hands me a note that reads, "I can't talk about the pigeons in front of Holly. I don't want her to know."

The pigeon bodies are stashed in the freezer for the next twenty-four hours to kill any parasites. The feet are thrown away. The wings are for the taking. They make good behavioral enrichment for the big cats, monkeys, and raccoons. In Nutrition, Amber Cavett collects a couple of pigeon wings, pokes holes in them with a knife, and threads strings through the holes. I follow her down to Samburu's cage. He hisses at us as we walk up to him. She's been training him to take food more calmly but making only marginal progress. While the caracal is closed off to one side of his cage, Cavett slips into the empty half and ties the wings to the top of the enclosure so they dangle like a mobile.

She steps out and lifts a gate. Samburu rockets into the other half of his cage. The raccoon next door has scaled the side of his enclosure to see what all the fuss is. Samburu turns his face up to the wings, sniffs, then shoots straight up about four feet, front legs extended, and bats the wings with both paws. They sway like piñatas. A few more lightning-quick jumps and bats, and he's got the wings down and one in his mouth. The caracal lies belly down on the floor as he devours the wing. The feathers fan absurdly out of his mouth, the bones cracking loudly between his teeth as the cycle of life spins on.

The next morning, as I walk in from the parking lot just shy of

7 A.M., the dark silhouette of a student shoots past me in the gloom. Her name is near the top of the alphabet on the roll call. The teaching zoo's lights color the morning mist lavender. Just down the front road, the heat lamps glow a hazy orange in Parrot Gardens. The birds squawk lazily. A low mooing emanates from deep within the zoo. Clarence slowly lifts his anvil-shaped head and looks my way. At the morning meeting inside Zoo 1, CVP stands and congratulates yesterday's pigeon pullers. "We're very proud of you," she says. There's talk of which vents should be open or closed in Primate Gardens, warnings to keep extension cords out of animals' reach, and an announcement that the garbage disposal in Nutrition is clogged. The meeting over, the students sleepily zip their hoodies, hunch their shoulders against a damp chill, and head into the early morning gloom.

I walk with Castaneda down the gravel back road to Big Carns. She's still upset that pulling a pigeon upset her. Castaneda is five foot eleven inches. She's been patronizingly called "a big girl." She used to be even bigger and keeps a photo of herself seventy-five pounds heavier taped inside her planner. Whatever her weight, she's never felt or acted girly. In her family "whatever happens, I'm the one who has to be okay with it. I even cleaned out my sister's litter box when it was like sludge," she tells me. Standing by the cougar boys, Sage and Spirit, who bound about boxing with each other, she pauses and then says, "Just think about it. How can you murder something with your own hands?" She shakes her head.

"We don't seem to have any poop today," VanHollebeke calls. She's in charge of the cleaning crew of first years in Big Carns this morning. A day without poop here is a holiday. Most everyone agrees that the big cats produce some of the nastiest-smelling guano. As students gingerly pick up log after log with tongs, they often have to pause and turn their heads. The stink is solid and it gathers at the back of your mouth. One first year even pulls his T-shirt up over his nose to fend off the stench.

The cages in this area form a large letter C. This is what Taj, Kissu, Sage and Sprit, Kiara, Savuti, and Legend call home. C.J. also lives here, as do two servals. In the middle is the arena, a roomy enclosure with logs, a den box, and platforms. Several of the animals take turns spending a day per week in the arena. It gives them a change of scenery, room to stretch, and a bed of grass to lounge on.

Cleaning in Big Carns is like working in a federal prison. Doors have to be carefully locked, then unlocked, then locked again. Nearly every morning an animal is moved or "switched," as they say, into the arena. This involves a sequence of opening and closing doors that looks simple enough, but if one little mistake is made, an animal or human could end up hurt or dead. That is why a staff member is always present. Today, Legend comes out and Kissu goes in. This requires Legend to pass Kiara's cage, where the wolf and the lioness will go at it through the chain-link. Though they have been neighbors for five years or so, given the chance to quarrel violently, they will, so Kiara has to be closed into her bedroom. That takes some chicken necks and some pleading, but eventually the lioness sashays to her den box.

The wolf returns to her cage, and a small army of students swarm into the empty arena to spruce it up for Kissu. Students carry in a basket filled with cheap colognes, spices, talcum powder, spray deodorant, mouthwash, and lotion. Like the pigeon wings, these scents are behavioral enrichment for the big cats—something to mix up their day and a chance to use their natural instincts. The students dab perfume here and sprinkle onion powder there. One of the spices makes Kissu so ecstatic she drools and rolls on her back, but no one is sure whether it's the roast beef seasoning or the poultry seasoning or maybe the steak seasoning, so they scatter them all. Students exit, locks are unlocked, doors opened and closed. Kissu bolts into the arena, lips parted, eyes bright, darting from log to stump to rock, breathing deeply.

Taj, the nine-year-old Bengal tiger, has a cage twice as big as the others with a view of the back road. Here, she can slink along an end of her cage, stalking students and other animals on leashed walks as they go by. She also has a pool, in which she lounges with her tail draped over the side. Anita Wischhusen gingerly pokes a piece of chicken neck through the bars to lure the tiger to one side of her cage so they can clean the other half. "Don't put your fingers through," Brenda Woodhouse warns. With a low rumble, Taj steps her way and snatches the neck with her large teeth from Wischhusen's grasp. The student shudders, squeezes her eyes closed, and giggles as if she's just been tickled. As Wischhusen steps back from the cage, she says to me, "It's hard not to talk to Taj. I just found out that you can chuff at her."

"Only if she chuffs at you first," Castaneda says, standing nearby.

"Has she chuffed at you?" Wischhusen asks.

"No," Castaneda answers.

"Well, you got Schmoo for Davis week," Wischhusen retorts.

This is a very sore point for Wischhusen just now; it's becoming the same for Castaneda. While the second years take an upcoming, weeklong field trip, the first years have been assigned to take care of various animals. The competition was intense, as usual, for Schmoo. Only the star students get the star animal.

Schmoo is the only marine mammal at EATM. She is also the most time-consuming animal at the teaching zoo. Her student trainers spend hours thawing her mackerel and squid. She is in most of the zoo's educational shows. Her long list of commands must be regularly rehearsed so she doesn't forget them. Her student trainers are forever cleaning or fixing her pool, which was meant for a backyard, not for a sea lion. Consequently, staff assigns Schmoo only to students with straight A's and perfect attendance, even if they are only temporary caretakers. That is because the sea lion takes so much work that students' grades often suffer; the students have to be in a position to let their grades slip without risk of flunking out.

About a dozen first years asked to be assigned to Schmoo for the week. Only four were chosen. Wischhusen, who does not have straight A's, was not one of them, and she's pissed. She suspects it's because she is in her forties. However, she did get Kiara, which is far from a booby prize. Castaneda, who has straight A's and nary a blemish on her attendance record, got Schmoo. This is one of the main reasons she has worked so hard: to win these kinds of plum assignments. But what she didn't expect were the resentful comments. Wischhusen's isn't the first, and Castaneda is getting tired of it.

"Well, you got Kiara," Castaneda counters as Wischhusen in her pink wellies walks off.

Gary Wilson stands by the front office watching a second year with almond-shaped eyes and hair down to her waist give Cyrano, a blue and gold macaw with a slightly crooked mouth, a pedicure. She saws away with an emery board on his front nails. Cyrano keeps trying to pinch her hand with his black beak. Wilson watches briefly, pokes around at his palm-sized computer, and then announces her grade: "I'll give you an 85." He and the student walk to the other end of the zoo, where she collects Bubbles, a young opossum with beady, glassine eyes and little, fluttering black ears. The second year hooks a leash on the opossum, and Bubbles trots alongside her like a puppy out for a walk. "She's doing fewer open-mouthed threats," the second year tells Wilson. He gives her a 95.

Time always flies at EATM. It is already midterm week for the second years. They have to demonstrate how well they have trained their animals so far this semester. The way it works is that

Wilson, his PDA in one hand, a small video camera in the other, roams the zoo, meeting students at various cages at appointed times. He nearly always runs late, so the students often wait nervously by their animals, warming them up or gauging their frame of mind. Wilson, considered a tough grader in his academic courses, is a softer touch in his training lab. He often gives the students the benefit of the doubt: animals have moods; they don't feel well sometimes; they get distracted by scents, sounds, and sights; they can be scared by the wind or a maintenance truck driving by. Still, students are embarrassed and frustrated when after weeks of training their animals appear to have amnesia or, worse, nip them during a midterm or final.

Wilson also considers the animal itself, how experienced it is. He expects the students to teach more ambitious behaviors to veterans of training such as Abbey or Nick. For her midterm, Amy Mohelnitzky trimmed Abbey's toenails, though the pooch will only let the second year do it when she's lying on her side. Wilson gives her a 100. Anderson demonstrates the progress she's making in teaching Nick to pull a cart. So far, Anderson's gotten the chubby horse to wear blinders and a halter. She attached two bamboo poles to his halter and trained him to pull those. Nick complies but he's tense. Anderson gets a 90.

For nervous prey animals like Starsky, Wilson has far lower expectations. During her midterm, VanHollebeke sits on the concrete ledge and holds out the elongated spoon filled with alfalfa pellets to the wary cavy hiding at the back of his cage. She gets a B. Wilson also considers the species. In terms of brain power, Wilson says the reptiles are the hardest to train. Marlowe Robertson, a second year with a fondness for reptiles, is the first to have made any headway with Happy, the alligator, having trained him to follow a target pole. On the flip side, primates have so much brain power, Wilson says, "They have all day to think about how to do the least amount of work for a treat."

At the far end of the zoo, Wilson finds Becki Brunelli waiting by Goblin. Brunelli has graduated from sticking her finger in Goblin's ear to prodding it with a spoon. She squats down by Goblin's cage and calls the baboon over. Goblin lopes over and on Brunelli's command presses her nose against her cage.

"Ear," Brunelli says.

Goblin turns her head sideways, bringing her ear closer to the cage. Brunelli slides the shiny handle through the bars and toward Goblin's ear. She misses her target slightly. Goblin's eyes widen and she jumps as if she's been goosed. "Sorry," Brunelli says to Goblin and then puts the spoon handle in the baboon's ear. Wilson gives her a 95, docking Brunelli five points for her poor aim.

Next, Wilson meets up with a petite, freckled second year by Friday the raccoon's cage. The second year steps inside the cage with a scale. As soon as she sets it down, Friday tries to jump on. "Station! station!" she repeats, but the raccoon wants on that scale. The second year has to block Friday with her arms while she readies the scale. She and Wilson crack up. Despite the raccoon's enthusiasm, she gets a 90: "His stationing needs to be more solid," Wilson says.

Wilson ends the morning at the hornbills' pen with the second year he considers to be one of the best trainers in her class, April Matott, a small, understated blonde from Albany, New York. Matott asks Wilhelmina, a thirty-year-old Abyssinian ground hornbill with a crippled wing, to step up on a perch. The ancient, unsteady bird complies. This is a major accomplishment because early in her training Wilhelmina fell and accidentally spiked herself on a shoot of bamboo. Matott gets a 100.

"There's only a few students who are training the animal and not vice versa," Wilson tells me.

* * *

There are many unexpected lessons at the teaching zoo. Students learn that the camel's breath smells like rotting, fermenting broth, that the owl's poop is stinky and sticky, that tortoises pass loud, stinky bursts of methane. "I've learned that everything farts," a first year from New York City tells me. EATM desensitizes these students to things that would make most people vomit. A squirrel monkey pooped on a student's head. "It was like apple sauce," she says. A first year tipped over a bucket of blood in Nutrition. It dumped into her boot. While another was organizing the walk-in freezer, a shelf full of frozen pigeon parts showered down on her. A frozen, headless pigeon body hit her in the mouth. Feathers stuck to her lips.

During the morning cleaning, a small contingent tiptoes into Zulu's cage, gingerly stepping over creamy pools of his ejaculate. A first year accidentally dipped her hand in one. In the Rat Room, students scan the bins looking for half-cannibalized rats to pluck out. While hosing poop out of cages, many a student has shot a stream hard at a log, only to have the water splash up into their mouths. And then there's Dr. Peddie.

No one can quite prepare you for the vet's ability to make your stomach turn. He lulls you with scientific terms, then bam! He hits you in the gut with a vivid description of an abscess. By their very nature his vet classes cover nauseating material, but he just seems to relish it.

Take this morning's lecture, for example. It starts out innocently enough as he describes neutrophils, lymphocytes, eosinophils, basophils, and monocytes to a class of second years. A few struggle to stay awake to note down what the various white blood cells do, their ponytails swaying as their heads droop. "It's hard to make blood interesting," the vet confesses, his brow knitting.

He cuts to a favorite topic, not exactly a crowd-pleaser, but one that is sure to wake a few sleepy heads—pus. He recounts the story of his best friend, who just had surgery; the incision became infected. Pus was just pouring out of it, he exclaims. "I don't like your stories," a student in back says. He flashes slides of sick animals: a puppy with lockjaw, a lamb that has to be put down. "Not all pus is liquid. This pus is like cheese," he says.

"Yuck," someone says, which prompts a chorus of shushes. An image of two large worms flashes. Someone moans, "Oh man."

Last but not least is the image of a goat with something indistinguishable but hugely swollen hanging off its behind. "You're looking at the south end," Dr. Peddie chortles, pointing at the slide. "That is his scrotum. It's as big as a melon. *This* is full of pus."

The next morning finds the vet far less jovial. Dr. Peddie has ruled today, the second Tuesday in October, Zeus's last. Dr. Peddie will put down the ailing iguana some time this morning, after the students are done cleaning the zoo and before class.

Down in the claustrophobic Reptile Room, where the snakes and lizards are stacked in cages one atop the other, the morning's cleaning is just wrapping up. "Everybody pooped today," Jena Anderson tells me proudly. "Everybody had stinky poop." Now they've moved on to feeding. Anderson attends her "barbecue." She's arranged pinkies and fuzzies, dead baby rats with black eyes bulging through their paper-thin eyelids, on a space heater like so many hot dogs. While the pinkies and fuzzies heat, Anderson heads over to Nutrition to make the turtles' breakfast. She loads produce onto a tray and then adds two roses dusted with cookie crumbs. "Presentation is everything," she says. "Bon appétit."

In addition to training Nick, Anderson is a Reptile Room manager. She and five other classmates oversee the general care of the teaching zoo's scaled collection. This includes several very large snakes: Precious, the yellow anaconda; Morty, the Burmese python;

and Ceylon, the Indian python. There are smaller snakes, such as Mupu II, the California Mountain king snake, and Kia and Kio, the sand boas. They recently produced a pile of wriggling baby sand boas. Conan, the popular Chinese water dragon, lives here, lounging in a pool shaped like a volcano. So does Howie, the blue-tongued skink owned by a movie trainer, who retrieves the lizard for a job occasionally, and Dot, the African gecko that doesn't stick.

In each class, there are typically only a handful of students interested in the reptiles. Usually, there are also a handful who'd rather go through life never having touched a snake. They'll have to get over that. Currently, only Happy, the alligator, is trained, but a good number of the reptiles are used in the teaching zoo's shows. A favorite stunt for private birthday shows is to get a group of volunteers from the audience, have the kids line up, close their eyes, and hold their arms out. Then the EATM students lug out Ceylon and drape her over the kids' outstretched arms. And when you clean the Reptile Room, there's a good chance you'll have to touch a snake or two. The large snakes like Precious are carried outside and allowed to slither around on the pavement for a stretch.

This morning in the Reptile Room, all the animals have been fed except for Zeus. No last meal. No last cleaning either. She sits near the back of her cage oblivious to her fate. Marlowe Robertson arrives looking forlorn. She is Happy's trainer and, like Anderson, a Reptile Room manager. She has on a big leather glove. She slides open the glass door to Zeus's cage and reaches in. Two more Reptile Room managers appear. They collect Blaze, the corn snake, and load him into a carrier. The snake has been very sluggish and may need to be euthanized as well. We start up the gravel back road, a large lizard's life weighing on our minds. Robertson carries Zeus on her upturned palm. The iguana remains still, her rounded nose forward. Robertson's eyes are wide and moist. No one talks, except that when we pass Wendell, the pygmy goat, standing atop his box, Robertson sings out in a wobbly goat voice, "Hello, Wendell." He flicks his black tassel of a tail. On we walk. The lethargic corn snake nearly wiggles out of its carrier in a burst of unexpected energy. By Parrot Gardens, a dark-eyed second year hurries up to us. "Can we say good-bye?" She strokes Zeus's head and stares into the lizard's unblinking eyes.

Dr. Peddie leans out of Zoo 1, tight-lipped, grim-looking, with rubber gloves on, and a syringe in one hand. He waves us in and quickly gets down to business. While Robertson holds her, Dr. Peddie injects a dose of pentobarbital sodium into Zeus's ribs. Suddenly, she is animated and trying to squirm out of Robertson's hold. This is typical of animals about to be put down, Dr. Peddie says. He has Robertson hold Zeus's belly down on the table in Zoo 1 so that the lizard's tail drapes over the side. Dr. Peddie gets down on one knee and injects another dose of the barbiturate into the base of the iguana, just under her tail. He gets lucky and hits a major vein.

"Reptiles have low blood pressure," he says. "This is going to take awhile." Robertson holds the lizard against her chest. Zeus grows stiller, but waves her head around occasionally. "Be careful because she can still bite you," Dr. Peddie warns. As death slowly comes for the lizard, the vet examines the corn snake, which seems to have miraculously recovered. Dr. Peddie sets the snake on the table, then rolls it over. The corn snake quickly rights about half of itself: not bad. Dr. Peddie gives the corn snake a reprieve. Meanwhile, the tip of Zeus's pink tongue lolls a bit. Then her eyes slide closed, but her sides still push in and out. Her plump tongue protrudes more and more.

Her whole tongue sticks out. Her stillness is indisputable. Robertson smothers a sob. Tears run down her cheeks. Dr. Peddie holds up a plastic garbage bag, and Robertson gently lowers the iguana inside. She wraps both her arms around the bag and clutches Zeus in a way she never could while the lizard lived. Holding the dead iguana close, Robertson walks out of the classroom into the constant California sunshine.

Later, after the bunnies have been bathed, after a student has nearly been knocked down by Sequoia while leashing the mule deer for her midterm (Wilson gave her an 80), after Rosie, the baboon, and Abbey have almost run into each other while out on walks, and after Dr. Peddie has lectured the first years on what he calls the wow of the heart, all the second years gather for one last look at Zeus in Zoo 1.

Her still body slides out of the green plastic bag onto the very same table where the iguana got her lethal injections. The students gather around. A single fly buzzes the room. The iguana's pink tongue flaps sloppily. The vet points out the papery layers of loose skin on Zeus. She'd begun to molt, he says. Her front legs are robust, but her back legs are withered and flaccid. Robertson stands next to Dr. Peddie. She looks on, interested but sad, as the vet turns the iguana on her back. He cuts a neat straight line right up Zeus's belly, exposing her glistening pink insides. There are yellow-orange globs of fat tucked around her three-chambered heart. Dr. Peddie runs his fingers along her intestine. "This is called slipping the guts," he says. "I feel something like tapioca pearls."

"Eew," someone moans.

"See, there is one hole for everything, pooping and giving birth," he adds. The only abnormality he finds is a small yellow cyst on the liver. He pinches the cyst and a discharge comes out. "It's cheesy," he says. Then he pokes Zeus's small heart with his scissors. The meaty pump seizes. That sets off a wail of disgust followed by hearty chuckles. Some students shake their hands, squirm, or roll their eyes.

"Do it again, Dr. Freakenstein," someone calls.

A mischievous grin on his face, Dr. Peddie pokes Zeus's heart again. It lurches, prompting more cries and laughs. Robertson, stone-faced, strokes the dead lizard's foot with her index finger and says a silent good-bye.

Briz

Bill Brisby appears to have had some kind of midlife crisis—a rather spectacular one. Like a solar flare, its effects spread far and wide and can still be felt. In the 1960s, he was an award-winning high school science teacher. He had a crew cut, geeky, thick-framed glasses, and was even a bit pudgy, according to his longtime friend and colleague Lynn Doria. He lived with his wife, Beverly, and three children. By the end of the 1970s he had morphed into an exotic animal trainer and started a one-of-a-kind school. He let his close-cropped hair grow beyond his ears and sprouted a shaggy, graying beard. He dressed in khaki, adopting that African big hunter look that so many trainers and zookeepers favor. He divorced his wife of twenty-six years and married one of his students thirty years his junior. "Everything about him changed," says Doria. "He went from one extreme to the other. He never found the middle."

Briz, as everyone still calls him, was the kind of man who is easily mythologized even while still alive. He had the bravado of a man who could wrestle with a lion—and did. He was a charismatic, inspiring teacher, the kind that could beat you down and build you up at the same time. Former students still recite Brizisms such as: "When you know everything, you know nothing," "It's okay not to know something, but go find it out," or my personal favorite: "You can train a chicken and do a great act in Vegas."

He had the sharp eye of an entrepreneur, spotting an opportunity from a far distance. He was audacious enough to teach himself how to train dangerous animals. He was a glass-half-full type who believed in the power of hard work and pushing yourself to the limit. He didn't need much sleep. His second and much younger wife had trouble keeping up with him. He had big hands and a big voice. He was a ruggedly handsome alpha male who did things his way, no matter who it pissed off. Women—particularly young women—were drawn to him.

Most everyone says that you either loved or hated Briz, which may be more myth than truth. I eventually realized this was a polite way of saying he could be a first-class asshole. I did come across a few alums who hated him, but even they had a deep, if grudging, respect for the man who made such a wild dream come true. "Some people thought he was arrogant," says Jamie LoVullo (EATM '83). "I thought he was incredible. Of course, he wasn't perfect."

There's not too much in Briz's early life that points to his later transformation. According to a lengthy obituary, Briz was born in 1924 and grew up in LA's Compton long before it became known for gang warfare and inner-city strife. His family had animals—nothing exotic, but plenty of cats, dogs, chickens, goats, and rabbits. Though he taught science most of his career, his B.S. was in physical education from Colorado A&M, according to college records. He embarked on what would appear to be a fairly ordinary life and career, coaching football and teaching at Fillmore High School, where he remained from 1950 to 1969. However, Briz wasn't ordinary. At home, his family had foxes, raccoons, a South

American tree otter, ducks, a ring-tailed cat, and rabbits. He became head of the science department. He won awards. He set up an outdoor classroom on Rincon Island, a man-made island off the coast near Santa Barbara. He convinced Edward Teller, the father of the H-bomb, and Charles Richter, who invented the earthquake scale, to speak to his science club. He invited exotic animal trainers Hubert Wells and Wally Ross to visit his classroom.

There are various myths about Briz. One is that in the 1960s he trained dolphins in the Navy's program at Point Mugu in Malibu. His obituary says he was the curator of the center's aquarium; however, Dr. Sam Ridgeway, who began the Navy's marine mammal program, says Briz never worked there. Ridgeway recalls Briz's coming to the lagoon for water samples, but that's it. Everyone assumes that Briz had a degree in science, but he followed his physical education degree with a master's in secondary administration, receiving that degree in 1969 from the University of Southern California. That same year he joined the faculty at the newly minted Moorpark College. Built on 134 acres of old ranch land in what was then primarily a farming town, the school opened in 1967 and its student body quickly swelled.

In 1971, Briz taught a course in wild animal training and management. He lectured on training and behavior one day a week and on another took the class to different facilities, such as the former MarineLand USA. A number of his students were offered jobs. That got him thinking. According to Doria, he and Wally Ross and Martine Colette, who now runs a sanctuary called the Wildlife WayStation, discussed setting up a school for trainers. When that didn't pan out, Briz proposed the idea to Moorpark. He suggested a new major, one that would prepare students for jobs in animal parks, zoos, aquariums, and even circuses. These were vocations no community college had prepared students for in the past. A new institution is usually a fairly open-minded one, and Briz was persuasive. Moorpark bit. "He hit the time when the emphasis was on adult ed," says Dr. Peddie. "EATM came into that. Moorpark started a program in aviation at the same time. A lot of these programs fell by the wayside, but EATM didn't."

It was 1974 before Briz had enough courses to constitute a major. Doria, who had a house full of mangy kittens, opossums nicked by lawn mowers, and birds with faulty wings, became the program's first graduate, Briz's girlfriend, and his accomplice. He called her Little Red. A small woman with prominent cheekbones and a mane of red hair, Doria had moved from her hometown of Chicago to Los Angeles to be an actor. As soon as she graduated, Doria joined the EATM staff. She retired in 2000.

The twosome began rounding up animals for the teaching zoo, starting with Kiska. They drove down to Van Nuys to pick up the eight-month-old wolf that they'd been told was tame. He was not; he had been raised with a dog that he romped with but was not used to people. They had to push the wolf into the back of Briz's truck, which was covered by a camper shell. The wolf hid for two days on the top shelf of the camper. Doria and Briz took turns sitting in the camper with him, trying to accustom the young wolf to their presence. "I wondered if this was a good idea," Doria says. "I would never do that again. I didn't know any better." At the time, Doria assumed Briz had worked with a wolf before. Years later, she says, "He told me he was faking it."

When the wolf finally emerged from the camper, they leashed him to a post during the day and housed him in a cage in a maintenance building at night. They named him Kiska. The first official class of students did their night watch with Kiska, then the teaching zoo's only animal. Before long the compound began to bulge. It was the loosey-goosey seventies in southern California, when you could buy a baby elephant from the local tire store. People dropped off unwanted, unusual pets. Briz, Dr. Peddie says, could never say no. The students built the pens and cages to keep up with the new arrivals. The original compound was small, about three-quarters of an acre, and dense. You could see from one end to the other. It sat at the far end of the football field on low ground that regularly flooded during heavy rains. Students would get in trouble for riding the camel on the well-groomed turf.

Briz and Doria drove to Long Beach to collect two neurotic pet capuchins, one of which, Koko, still lives at the teaching zoo. They put them in a cage atop the station wagon and zoomed back up the highway. Briz got word that a safari park in Irvine had a lion cub they didn't want—a male with a chunk bitten out of his ear. Briz, Doria, and some students loaded the cub, which turned out to be a female the size of a Great Dane, into the back of a station wagon. They named her Carol. Briz and Doria had often snuck animals into the zoo, unbeknownst to the college honchos. There was no hiding the lion. Administrators asked Briz how he would pay for the upkeep of the king of the jungle and the rest of his expanding zoo. With educational shows, Briz answered. EATM itself would pay for most of the teaching zoo's upkeep, as it still does. That appeased the administrators. Briz, untethered from the college's purse strings, had even more freedom to run his kingdom as he saw fit.

It was *his* kingdom: Briz picked the animals and picked the students. He ran the program like a military school—an odd choice for a man who told Doria he had washed out of the service after an emotional breakdown. At EATM, he was the brass and nothing made the point clearer than the pipes. They were neatly stacked at one end of the zoo. They were sewer-sized, maybe twenty feet long. They were rusting metal. It took five or six people to lift just one. During orientation week, Briz would order the new first years to move the piles of pipes to the other end of the zoo. If they dared to ask why, the answer was a gruff "Because I say so." It was hot. They got blisters. The blisters broke on the pipes. Briz yelled at them, calling them "stupid broads." After the pipes were neatly stacked in their new location, they would sit there rusting until the following fall when the next new class arrived. Then the new students would lug the pipes back to the other end of the compound, and so it went year after year.

"He was a total dictator, also to his staff," says Dr. Peddie. "I was one of the few people he did not lock horns with. Not that the students didn't love him. They feared him."

"If he told you to jump off a cliff, you might have," says Dorothy Belanger, a senior keeper at the Los Angeles Zoo, who graduated from the program and now works there part time.

If you wanted democracy you came to the wrong place. Briz didn't suffer anyone questioning his authority, which rankled some students. One alum told me he was more a bully than a leader, one who favored pretty girls and loaded his tests with trick questions. If he didn't like you, he'd make it clear. His big, booming voice could be heard the length of the compound. He believed in the crucible approach. When Susan Cox was a student in 1985, she asked Briz if she could quit working with the parrots because they wouldn't stop biting her, tearing her hands to shreds. No was the answer. When Diane Cahill, Briz's second wife, was a student, she hid her snake phobia from him, knowing if he found out, he'd assign her to one.

Students were required to be at the school at 6 A.M. every day of the week from September until June. Every third or fourth night they had to stay over, sleeping on hay in a barn or in with the reptiles, which occasionally got out. "We worked, worked, worked," LoVullo says. Kris Romero, who graduated in '88 and teaches part time at the school, remembers being so exhausted that she hallucinated seeing ants all over a countertop. If you got three tardies, you were out. Briz counted tardies as being late for or missing one class. If you missed a morning with three classes, bingo! You hit your limit; off you went. There was no grievance procedure. One alum told me of students setting up other students for a third tardy so that they would be expelled. "I saw some good animal people get kicked out," says Cox.

Briz's point was that the animals come first. He knew he couldn't make his students animal experts, but he could instill in them an exceptional work ethic for a field that requires utter devotion and long hours. "He wanted you to be married, like a nun would be married to Jesus," Doria says. Gary Wilson says students obliged because it created an esprit de corps. "Why do people become a Navy Seal and go through that hell?" he asks. "It's because it's an elite. It's an accomplishment."

Typically in these years, only half a class made it to graduation, what with students getting kicked out and dropping out. When Wilson went, his class started with thirty; fifteen made it across the finish line. When Brenda Woodhouse (EATM '80) graduated, only sixteen out of her class of forty were on hand. Many of the casualties were men. The college pressed Briz to accept more male students because there were so few and those few were often thrown out. "He was at odds with men," says Alan Kordowski (EATM '76), curator of mammal training at SeaWorld in Orlando. "So we didn't get along."

If Briz had trouble with men, he'd created the perfect world for himself. From the get-go, EATM attracted female students. Long hours of endless cleaning never discourages women the way it does men. It was a gender imbalance that proved a great temptation. "It was Candyland," says Doria. After his divorce, Briz became engaged to Doria, who was fifteen years his junior. Then he broke it off to marry Cahill, who was thirty years his junior. Dr. Peddie says Briz was a sexual athlete. "His students would give him anything. I think many of his students did."

The largely female student body combined with Briz's military style of running the school made for an odd mix that continues today. As many sports coaches know, most young men need to be knocked down a peg or two, but most young women need the reverse. Females do not generally respond well to being beaten down, criticized, and subjected to ironclad rules. Kordowski says

Briz did it to toughen them up. "Back then women were in the industry, but all the bosses were men."

Cox says Briz told them they could expect to have to sleep with their bosses. He would say things like "Come here, little girl." "He was harsh. He was brutal. He was sarcastic," she says. She's grateful for it. She went to EATM when she was thirty-six, married, and had two kids. She had grown up a latchkey kid in a family that taught her to think small. She married young and let her husband run her life. Without Briz and EATM, she says, "I would not be who I am today. I'd still be afraid to drive on freeways, afraid to go places by myself. I wouldn't have done any of the things I've done."

LoVullo, who is a rep on movie sets for the American Humane Association, was, as she admits, "a really horrible teenager." She snuck boyfriends into her house. She ran away. She wanted to drop out of school. Then she heard of EATM. A competitive horseback rider, LoVullo could envision a future working with animals and had a reason to stick out high school. Animals and EATM got her on the right track, and she mostly credits Briz's tough love. "I'll never say a bad word about him." After graduation, LoVullo kept in touch with Briz. She found that once she was no longer an EATM student, Briz was a bit of a teddy bear. "He played the part of a hard-ass," she says. "That's not really who he was. That is what he did to make better students."

For many of the alums from the Briz era, these were the glory days of EATM, a golden, not-so-distant past when the school ran as it should. The students would trek up to nearby Carlisle Canyon to take training classes with Cheryl Shawver of Animal Actors of Hollywood. The company's owner, Hubert Wells, the legendary movie trainer with a suave Hungarian accent, kept some of his animals at the teaching zoo, including the very first Schmoo.

Though Wells didn't teach, he'd give a talk here and there. While a student, Cahill helped Wells raise a pack of wolves for the film *Never Cry Wolf*. LoVullo remembers Wells arriving and opening the back of his truck to let three lions out. They trotted alongside him off leash. "He was a god to us," she says. Wally Ross, the circus man and movie trainer, was about as well. When EATM briefly took in a young elephant that a local group of Hare Krishna couldn't handle, Ross supervised its student handlers. Frank Inn lumbered through the gate, then an old and very fat man, and put Benji through his paces to wow the students. He even gave Briz a look-alike pooch.

It was not such a shining era, though, for many of the college's administrators. A stickler for his own rules, Briz repeatedly flouted the college's. Briz pretty much ran EATM as he pleased. He was bounced from exasperated dean to exasperated dean. "I don't know if rules applied in his mind," says Brenda Woodhouse. "He had majorly pissed off the administration. No one wanted to deal with him."

When Briz retired in 1985 at sixty, the college named Wilson, then a part-time instructor, the new EATM director. Briz chose him as his successor. Doria had hoped to take over but she didn't have a graduate degree. Wilson did. It was an opportunity of a lifetime but an impossible task. Briz's shadow loomed large. It didn't help that Briz stayed around another year to teach one last class as Wilson tried to take charge. Briz was larger than life; Wilson was—is—not. Where Briz boomed, Wilson talked. Like an old-time preacher, Briz had struck the students with a mix of fear and awe. Wilson was young and approachable, a scientist at heart. Where Briz was obviously passionate and outspoken, Wilson is understated, and his ever-present smile is hard to read at times.

Wilson set about making amends with college administrators and making the program less of an outcast. He beefed it up academically and he codified the rules Briz had carried in his head, putting them to paper in a handbook. Alums and staff began to worry that EATM was losing its edge, that it wouldn't survive without Briz. Briz voiced his displeasure with Wilson's changes to the students, undercutting the new director's authority. "There was pressure from the college to make the students come first, the animals come second," Doria says. "[Gary] was more for the students. He was gentler. He was trying to be the softer side of EATM." But then anyone would look like a softy compared to Briz.

Nearly two decades since Briz retired, the hand-wringing over EATM continues. Alums still complain that the program is no longer tough enough, even those students who studied there when Wilson, the alleged softy, was director from 1985 to 2001. They say the school is no longer the gold standard, that the graduates are arrogant, demanding, and even whiny. "They are stripping away everything which made it special," says Netta Banks, a rep for the American Humane Association's Film & TV Unit.

The complaints have become even louder since Dr. Peddie took charge in 2001, though the alums don't blame the vet but rather the college administration. Brenda Shubert, who became EATM's dean in 1999, has brought the program into the fold, making it less a private school within a public college. In so doing, she changed EATM requirements that essentially weren't legal at a community college in California, a necessity in an age when people are not shy about going to court. For example, students are no longer required to be at the zoo seven days a week, though many still come that often. The old rule made the students spend hours at the zoo for which they received no course credits. This was not kosher. Now they are broken into two groups: one cleans Sunday through Wednesday; the other cleans Wednesday through Saturday. The cleaning is now part of a course called Zoo Skills, for which they get three credits.

The alums don't like this, but what really bothers them is that night watches are no more. Previously, students had stayed overnight on the compound with no faculty supervision. The college ruled that practice a liability bomb ready to go off. Night watch, says EATM staffer Mara Rodriguez, was "what made it real. That you were going to become a wild animal trainer." People speak of it reverently, as some kind of baptism. There was even an EATM superstition that the animals spoke at midnight on Christmas Eve. Mark Forbes of the studio training company Birds & Animals Unlimited is one of the few EATM alums I spoke to who downplayed night watch. "We'd go around and wake animals up," he says. "What are you going to do if an animal is in trouble? It was just to make it hard. It wasn't necessarily good for the animals, nor did it make you a better trainer."

What Forbes doesn't like is the lottery. In the past, the application process to EATM was long and involved, similar to a graduate program. You needed three letters of recommendation and you had to have work experience with exotic animals already. Of the 400 or so who applied, 100 or more were interviewed by the staff, who then chose the incoming class of about 50. Moorpark College is required by law to accept any student with a high school degree or equivalent. That EATM was picking and choosing its students was illegal for all those years. Beginning with the class of 2003, students are selected by lottery. They must have completed five prerequisites, such as freshman English and introductory biology, and they must attend a meeting for prospective students at the teaching zoo in April or May. This policy has greatly reduced the pool of applicants to about 100. Now, your chance of getting into the program, according to EATM's Web site, is about 1 in 2. Incoming students, however, can no longer claim a certain cachet. Getting into EATM is no longer proof that you are exceptional, just that you are lucky. So far, lady luck, the staff says, appears to be on their side. Dr. Peddie says the students selected by lottery are just as good. His beloved second years—one of the best classes ever—were chosen by chance. This does not allay worries, though. "At Moorpark they used to be able to guarantee they'd gotten the best," says Gary Priest, the curator of behavior at the San Diego Zoo and Wild Animal Park. "Now they can't."

What if you take a glass-half-full approach to EATM? The physical demands may have waned, but the academic demands have waxed dramatically. Now that the students are not so physically exhausted, they can study more. There are eight 4.0s currently. The attrition rate has dropped significantly, from 50 percent to 10 percent or less. Students also get more experience with the animals. Under Briz they generally worked with one animal per semester. Now they care for a minimum of four animals and train two.

The animals have it better. The compound, which was moved to the top of the ridge in 1990, is a big improvement over the cramped quarters of the original. In 2002, a staff position was created specifically to manage the teaching zoo. Michlyn Hines (EATM '84) was hired away from the Los Angeles Zoo, where she had worked for seventeen years. Since joining the staff, she has set about making the teaching zoo more professional, including plans to apply for accreditation with the American Zoo and Aquarium Association (AZA). Accreditation will raise the teaching zoo's status and make it possible for other zoos to lend animals to EATM. Plans are in the works for EATM to get a brand-new building; the college has approved $7.9 million for a new two-story building that will include classrooms, a dining room, a box office, and even a gift shop. If all goes according to plan, construction will begin in 2006.

So why don't alums notice these changes? Rodriguez says that the more the school changes "the less they can relate to it. There's a fear in that. They don't know how it works anymore." Mark Forbes says alums romanticize how hard the program was. "That way we can feel what we did had meaning. There wasn't a point to working that hard. Zookeepers don't work that hard." I also began to wonder if it isn't just human nature, that urge to lord it over younger generations because you had it tough. Or maybe it's animal people just being animal people. The truth is that the way Briz ran the program would be impossible today. It should have been impossible then—a teacher at a public school tossing students out at whim, making them work slave hours, sleeping with the students, and calling them "broads." These excesses have long since been huge no-no's on college campuses. Briz is lucky he got out when he did, before some student thought to take him to court.

It's also lucky Briz came along when he did, at a time when John Wayne types could still get things done in their big personality way without setting off an avalanche of grievances and court cases. In this sensitive, liability-obsessed age, it's hard to imagine anyone creating anything as outrageous as EATM or a renegade on the scale of Briz wrestling with lions in the hallowed halls of academia. For better or worse, the Brizes of the world are an endangered species.

After he retired, Briz moved with Cahill and their daughter to Camp Nelson in the Sequoia National Forest. There he morphed a third time, this time into the Owl Man. As a science teacher at the Clemmie Gill School of Science and Conservation, Briz donned a suit of buckskin and lectured to fifth graders while holding a horned owl on one arm. He'd become a kind of aging man of the mountains, a thinking man's Paul Bunyan. He was still a rule breaker. He took Sammy, his Lhasa apso with a horrible underbite, with him to work, though it wasn't allowed. He'd put the little dog out on his nature trail and use him as an example of a vertebrate.

He and Cahill divorced in 1988. By 1991, Parkinson's disease and prostate cancer had made Briz retire for good. Eventually, his son moved him to a retirement home, where, according to his obituary, he watched football and read Louis L'Amour westerns. On his front door he hung a photo of himself with his favorite lion, Chad. When Doria visited him, they'd talk about EATM. He told her he'd rather see it closed than come to what it had. An EATM alum, Gary Mui, interviewed him there for a film he was making about the school. Cox went with Mui. Briz was feeble, she says, though he still had "that twinkle." In the film, Briz's voice is weak. He looks at the camera through big glasses and describes in a low mumble how he used to irk his family during hikes by never letting them sit down when they were tired, making them push on just a bit and then a bit more. "You can always go a little farther," he murmurs. "You can get a lot of things done you didn't think you could."

On New Year's Eve 2000, Doria called him on the phone. They chatted briefly. He said he was tired. Briz died the next morning. He was seventy-six. He couldn't go any further.

The Fire

The third week of October the first years get the zoo to themselves at last. The second years pile into three vans and head north for the traditional whirlwind field trip referred to as Davis week. Like new mothers hiring babysitters, they leave long lists of dos and don'ts for the first yeas. They drive off fretting that their animals will not survive without them. These worries, however, recede as the second years ask for behaviors from an elephant at Six Flags Marine World in Vallejo, play with the puppies at Guide Dogs for the Blind in San Rafael, and feed giraffes at Safari West in Santa Rosa.

Back at the zoo, the first years get at least a little of what they have longed for—animal interaction. They chat with Louie, the geriatric prairie dog. They tender chicken necks through the bars to cougars. However, it is not the lovefest they dream of. The first years can only interact with the animals they have been assigned to, and the second years have set strict limits on that. Mostly, first years are allowed only to say "Hi" and "Bye" to their charges—no baby talk, no gossiping, no existential conversations. The first years assigned to Rosie, the baboon, are allowed two additional words: "Good girl." The ones working with Rio can ask the blue and gold macaw for behaviors inside her cage but cannot pick her up. Wischhusen gets more interaction than most. When she tells Kiara to go into her den box, the lion thunders right in her face. "The power of her roar exhilarated me," she says.

The first years assigned to Schmoo are forbidden to chat up the sea lion and must keep to perfunctory greetings. Nor may they ask her for behaviors—nary a wave. This seems to put Schmoo in a mood. The sea lion is typically as busy as a socialite. With her second year trainers gone, the sea lion's schedule of shows and training sessions comes to a standstill. She sinks into a funk deeper than her pool and refuses to eat.

At 7 A.M., 9 A.M., 12 noon, and 4 P.M., one of the four temporary Schmoo girls appears at the sea lion's cage with a plastic container of smelt, mackerel, and squid carefully prepared for her highness. They usually leave with what they came with—a full container. Each time Linda Castaneda steps close to the fence, Schmoo jumps out of her pool and waddles over to see who's there. "Hi, Schmooooo," Castaneda coos. Nearly blind and somewhat deaf, Schmoo takes a beat or two, nose upraised, whiskers twittering, head cocked, and then she realizes the silhouette is not one of her usual trainers. As Castaneda dangles a mackerel by its tail before her, Schmoo turns up her nose and closes her mouth tight. We are a species that takes things personally, especially a refused morsel. Snubbed, the star student sighs, frowns, and turns away from the star animal. For Castaneda what was an honor quickly becomes a royal pain in the neck. She's taken to calling Schmoo a "punk" or a "brat." She does not savor the prospect of telling the second years that they couldn't get Schmoo to eat.

To Schmoo, Castaneda and the three other first years must seem very dull. They want to give her fish for free—no barks, no spins—but Schmoo is used to working for her food. Handouts just won't do for this prima donna. The hunger strike is doubly worrisome: no fish means no meds for the aging sea lion. Schmoo has done this in the past when previous second-year trainers have left her in the care of first years. Because this is a teaching zoo, the staff wants the first years to figure Schmoo out and to tempt her appetite. The learning experience, though, cannot go on indefinitely or the cost of the lesson could be one dead epileptic sea lion.

During Davis week temperatures top 100° every day, though it's late October. The air in Hoofstock grows thick as the plentiful camel and sheep poop bakes in the sun. The doors are swung open on the Reptile Room. The big cats snooze the day away, rousing to swat at the flies that buzz their ears. The Santa Ana winds gust southwest, rasping the already withered landscape; there's been no meaningful rain since June. All the foothills surrounding the zoo have faded to dun brown under a garishly blue sky. Brush fires have broken out across Southern California, including a sizable one in Burbank, where the palm trees are reported to be burning "like tiki torches."

The crisped terrain around Moorpark may look like it could self-combust, but wildfires in Burbank seem a distant threat. Each fire season there's some smoke on the horizon here, even flames, but nothing very serious. It's earthquakes that people worry about around here.

As the week wears on, three things remain constant: Schmoo won't eat, the Santa Anas won't let up, and the mercury won't dip. The staff steps in with Schmoo and takes over feeding the sea lion. They can ask her for behaviors in exchange for food. Schmoo quickly gets her appetite back. The temporary Schmoo girls breathe a sigh of relief but feel defeated, especially Castaneda.

The class suffers its first casualty, a first year who has been repeatedly late for the morning cleaning. She came at noon one day and forgot to feed the sheep once. Some first years think she got what she deserved, but others come to her defense. With the second years gone, no bosses and no snitches, a bit of anarchy breaks out amid the first years. They yak away with their animals. One first year ignores her second year's diet orders for Clarence and starts stuffing the Galapagos tortoise with produce. The heat and the responsibility begin to fray people's nerves.

Weather forecasters promise cooler temps are on the way, but the thermometer registers 106° on Wednesday. On Thursday wild fires spark north of Moorpark in Piru, still well away from the teaching zoo on the far side of a river and a mountain range. Then on Friday a new line of fires ignites to the northeast in Val Verde near Santa Clarita. These blazes, too, are miles off on the other side of mountains, so no one worries at the zoo. However, the staff and first years can't help noticing that gray smoke like a low, dirty cloud peeks over the mountains' jagged ridge. What they don't know is that this plume is so big it's visible from space.

On Saturday the Santa Anas drive the faraway blaze and its murk west. In Moorpark the smoke thickens to a gauzy blanket. Like a smoldering coal, the sun glows red through the haze. The smell of a bonfire spices the air. In Big Carns, the cats pace peevishly and lift their noses into the sooty air. During morning cleaning in the Reptile Room the animals are antsy, odd: when Ceylon is set out on the blacktop to stretch, the Indian python tries to slither away; Precious, the yellow anaconda, does the same; Morty, the Burmese python with fangs bared, strikes at the plastic door to his cage. The zoo is open to the public on Saturday, but families arrive only to cut their visits short. The air is thick, and the temperature is well over 100° again. Residents near the college hose down their roofs and yards just in case. As the smog waxes and wanes, the first years start asking, "What if?"

Mara Rodriguez, the only staff person at the teaching zoo, tells them not to worry about a fire. She's seen smoke like this before. Rodriguez is in her mid-thirties. Though she doesn't, she looks like she spends a lot of time at the beach. Her shoulder length hair is streaked blond, and she has a tan. Hanging next to a couple of to-go menus on her office door is a quotation from Maya Angelou: "If you don't like something, change it. If you can't change it, change your attitude. No matter what, don't complain."

Rodriguez is an anomaly among animal people because she isn't one: "I'm passionate, but not like the rest of them, all these people goo-goo over animals. I've never been like that." Growing up, she didn't go to horse camp. She didn't lacquer her bedroom walls with tiger posters. She didn't even feed the family dog. "I'm a people person," she says. "I didn't grow up singing to my chickens. I was a city girl who went to the mall." She was, however, taken to SeaWorld as a kid, and that's what got her thinking about EATM. She was accepted in 1988. "When I got here I finally found something I was good at," she says. "It came easier to me than anything had."

After graduation, she got a job at Animal Actors of Hollywood. Rodriguez worked long hours on the set but made good money. She even ended up on camera. For *House Arrest* she clung to an air conditioner as rats ran down her arms and back. On *Full House* she doubled for a cast member, letting a dog lick chicken baby food off her face. "I've done everything. I'm the luckiest person in the world," she says.

At EATM Rodriguez weighed 190 pounds but as a movie trainer she slimmed down, thanks to a penchant for laxatives and puking. Eventually, she got over her eating disorder, but it took years. Now she keeps a keen eye on the EATM students, watching for telltale signs, such as girls who touch their bodies a lot or go from a size 8 to a 2. She has her eye on two blond second years who are just too thin.

Rodriguez thinks of herself as a role model for the students. She can show them that a regular person, even a girly girl, can work in this macho field. Even though she regularly walks the cougar brothers on chains around the zoo, Rodriguez keeps her nails long and impeccably manicured. Her lips are always shiny with lip gloss. "I don't want to become the stereotype, someone who is always dirty, lives in a trailer, and wears baggy clothes," she says. "Some of these girls try to be very masculine. I don't mind asking a guy to move something for me." She's also a single mom in a field where plenty of women still don't have children, especially studio trainers. Her son, Noah, is five. She'd rather be with him than do anything else. "Animals are important, but not the most important."

Around 3 P.M., about the time the flames in Val Verde leapfrog highway 126 and hop west toward Moorpark, white ash as thick as snow begins falling at the zoo. Sitting by Clarence's enclosure, Anita Wischhusen listens to cinders like snowflakes softly pelt the tree leaves. With the second years due back late tonight, today is the last day the first years have the teaching zoo to themselves. Many of them bring their parents up to meet the animals they have briefly cared for. Wischhusen's girlfriend takes a picture of her feeding Kiara.

Around 4 P.M., as usual, the students start closing the zoo for the night. Castaneda gets in her car and heads for home in the San Fernando valley. Wischhusen and her girlfriend sit on the back of her pickup truck in the parking lot watching the smoke, now as dark as a thunderhead, roil to the east. Though the sun dips near the horizon, it is still 100°. Sara Stresky, a student with pale skin and a broad forehead from West Virginia, feeds Chui, the white-nosed coati, and Buttercup, the badger. She takes one last walk around the zoo to snap photos of the smoke and then leaves. Stresky doesn't get far down the road when she sees flames cresting the hills very near the college. She calls Rodriguez on her cell phone and turns her car around.

Rodriguez, who not only hoped for the best all day but also expected it, walks out the front gate to see for herself. The sky is a sooty black, and a huddle of police cars and fire trucks are in the college's parking lot. She calls for students to start putting together crates, just in case. Then, before she steps back into the teaching zoo, a friend drives up. Get out, her friend tells her; the fires are closer than you think. The police turn on their sirens and over loudspeakers order people to evacuate.

Behind Rodriguez, 150 or so animals sit in their cages. There aren't enough crates or people to move them. The ten or so first years still at the zoo have never handled most of the animals. The second years have all the school's vans. There is no emergency plan for a fire. It's up to Rodriguez to come up with one this second. As the sky flashes orange through swirling smoke, Rodriguez turns on her heel and attempts the impossible.

One student is sent to the phone to call for help and reaches several staff members, who set off in their cars toward the zoo. Castaneda, reached by cell phone in her car, turns around. Flames now crackle along the highway. She bursts into tears. At the school, Wischhusen runs for wet towels and face masks. A few students construct crates in front of Zoo 1. The rest dash to Parrot Gardens to pack up the birds first because the smoke will smother their delicate lungs. Stresky charges into Banjo's cage. The macaw chomps her arms and hands over and over each time she reaches for him. Chieftain nails another first year in the stomach as she tries to grab him. Cookie, a loquacious African gray, goes into a crate—no problem. In the mayhem, Bwana, a turaco, flies off. Somehow, all the macaws are crated.

Stresky grabs a falconer's glove and runs to the Mews to get Laramie, the golden eagle. He reaches with one taloned foot toward her gloved hand but won't step up with the other. Without a hood on he won't climb on her glove, but Stresky doesn't know that. Finally, the eagle jumps to the floor, and Stresky herds him into a crate. She chases a raccoon endlessly around its cage trying to get the critter crated, gives up, and runs to the bobcat's cage to help another student. The cat hisses and bats at them. They back off, crate the serval next door, and return to the bobcat. This time he complies. Then Stresky hears Abbey's emphatic barks. I can't believe no one got her, she thinks, as she runs to the far end of the teaching zoo. She opens the cage door, and the briard mix bounds out and follows Stresky up the front road as if this were a great game. Stresky opens a truck door, and Abbey jumps happily onto the seat next to the crated bobcat.

Hudson, the beaver, is snatched from his pool. Helicopters flutter overhead. Adrenaline crests. Someone yells to keep calm. Students' tears pool atop their face masks. A first year assigned to Rosie, the baboon, for the week summons his courage and steps into her cage. He asks her to leash on. The baboon coolly does as told, even pausing at the door to her cage, as she's been trained to do. Nearby, a first year does something VanHollebeke has spent months hoping to do: she touches the cavy. She dives at Starsky, grabs him by the back legs, and tosses him into a crate.

Rodriguez, having rounded up and crated three squirrel monkeys, realizes that the biggest danger now is the pandemonium. She orders hysterical students to leave and keeps her own fears that the flames have them trapped to herself. While the first years worry about the animals' fate, Rodriguez worries about her own life and her five-year-old son and thinks, I don't want to die for the animals.

Michlyn Hines and her husband speed along back roads as

trees kindle like firewood and flickering embers drift by. They keep running into road blocks where they backtrack or beg to be let through. Her cell phone rings, and a first year says through hysterical tears, "We're evacuating the zoo. We don't have all the animals." The line goes dead.

Somehow, staff members Kris Romero and Holly Tumas make it to the teaching zoo, but with only moments to spare. Both women head for their favorite animal. Tumas scoops up a baby gibbon from Primate Gardens. Romero runs for her beloved turkey vulture, Puppy. When she was a student, she napped in the grass with the bird at her side. She let him pluck food from her teeth with his beak. As Romero drags a huge crate into his mews, Puppy flies out the door and up onto a fence post. Romero knows he can't out-fly the fire. She climbs the post and reaches for the vulture. He bites her outstretched arms over and over until she finally gets ahold of him.

A first year, noticing Olive still in her cage, opens the door. The baboon sits by the doorway as she is supposed to, but when the first year tries to leash her, the baboon threatens her. The student starts up the back road with Olive following. She leads the baboon into Zoo 1, closes the door, and gets Rodriguez. When Rodriguez sees Olive sitting on a desk, she is shocked that the baboon is not leashed. Now there are two problems: the fire and a loose baboon. Rodriguez can't leave her in Zoo 1. What if someone comes and opens the door not knowing there is a baboon inside? Rodriguez grabs a bunch of grapes off a desk to lure Olive into a crate set in the classroom doorway. The baboon screams at her and jumps over the crate and outside. Olive lopes toward Parrot Gardens while visions of the baboon running off into the surrounding neighborhood flash through Rodriguez's mind. Just then, Olive turns back and ambles into the crate.

Another student walks Nick to the front of the zoo, but the miniature horse is not small enough to fit in anybody's car.

Rodriguez leads Nick back to Parrot Gardens and puts the horse in a cage there. She finds Julietta, the emu she hand raised, loose. Though she's terrified of dying herself, Rodriguez can't leave the big dumb bird like this. She shepherds the emu into another parrot cage.

Over the loudspeaker the police announce, "If you do not go now we are not responsible for your lives." Students dump the contents of their cars onto the parking lot to make room for the animals. Wischhusen packs her truck with squirrel monkeys and birds. Castaneda tosses everything out of her car except her Diversity notes—there is a test next week—and loads four parrots, one squirrel monkey, and George, the fennec fox. Students sprint through the zoo turning on faucets and sprinklers, looking into the faces of Zulu, the mandrill baboon, Taj, the Bengal tiger, Savuti, the hyena, Kaleb, the camel, even Schmoo to say a quiet "I'm sorry. Good-bye." They open the door to the aviary so the plover, pheasant, wood duck, and turaco can fly away. Then they drive off into the smoke, darkness, and flames.

By now the blaze burns to three sides of the college. Since 3 P.M. when Wischhusen sat listening to the ash fall, the fire, like a fast approaching army, has marched ten miles west and encircled Moorpark. It burns wherever they look. The caravan stops briefly at a nearby burrito shop, but the flames are still too close, so they push on to the vast Target parking lot near the highway, hoping the sea of blacktop will protect them. They set up camp away from curious onlookers and between the back of the store and a large retaining wall. Students stampede into Target and take towels, bottled water, and anything that will make the fifty or so animals more comfortable. Crates are rearranged so that the birds are moved into cars that have air conditioning. Some students leave to rescue their pets at home. The rest wait, cry, and wonder if the teaching zoo is burning.

Hines and her husband finally arrive at the teaching zoo's front gate. She calls down to Target to tell them it is unscathed. Tumas and Rodriguez return. Other alumni and staff arrive. In about forty minutes another thirty or so animals are hurriedly crated, including the wolf, and evacuated. Dr. Peddie, who, like Hines, navigated a maze of back roads and police blockades, catches up with the makeshift zoo at Target. Everyone settles in, prepared to spend the night. They set up a first aid station and begin rounds, checking the animals in their crates. They keep a wary eye on the flames that color the night an ominous orange.

After a harrowing ride, the vans of second years stream into the Target parking lot at about 8:30. They bound out and frantically look for their animals. There is plenty of hugging and crying, but some second years shoo the first years away from the crates, while others don't think to ask the first years if they are okay. One second year announces that they may no longer talk to the animals.

The hill behind Target sparks. The EATM staff decides to return to the teaching zoo, where there is food and water for the animals and no curious public. The caravan winds its way back through the thick dark. Back at the zoo, the animals that have been crated are moved into the school's vans with the air conditioning running. The vans are parked in a single line pointing toward the front gate, ready to stream out at a moment's notice.

As the fire devours the sage brush, cactus, and paltry trees on the hills around the teaching zoo, any remaining animal that can be crated is. Samburu, the ornery caracal, is lassoed with a catch pole. The snakes and lizards are put in anything that can hold them, whether it be a trash can or a Tupperware container. Trailers are put on standby in Hoofstock for the animals there.

Brunelli finds Goblin on a high shelf in her cage, a blanket pulled over her head. The baboon must be sedated so she can be crated. Brunelli and another student tell Goblin to go to her bedroom, the alcove off her cage, where Dr. Peddie can more easily dart her. Brunelli tempts her with a watermelon chunk. Nothing doing. Brunelli puts stuffed animals in the bedroom. Goblin refuses. The baboon screams. She defecates all over her cage. Time is running out. Dr. Peddie takes aim at a moving target, the frantic baboon. Goblin soon swoons. Amid the mayhem, the crackling, and the smoke, Brunelli gently lifts Goblin and is struck by how light the baboon is while lowering her into a crate.

Dr. Peddie does not have enough anesthesia to knock down all the truly dangerous animals and, in general, he'd rather not. Anesthetizing an animal is a very risky business, especially when you don't know the animal's weight. If you guess just slightly wrong, you could have a dead lion on your hands. The big cats and Savuti are put in their concrete den boxes, which are then covered with wet blankets. Anything flammable is removed from their cages. If they have to evacuate again, the big carnivores as well as Zulu the mandrill will remain behind once more.

The students settle in for a night watch. The second years, back in charge, send the first years home. At about 2 A.M., the flames reach to the stars, and the sky pulsates and roars. The fire slides down the hill immediately west of the college like a lava flow. If the winds shift, the blaze may lunge at the school. The staff and students, people who spend their days gauging a lion's mood or anticipating a capuchin's next move, watch helplessly as this hungry beast rages through the night before their eyes. Operant conditioning holds no sway with this creature.

* * *

The wind does not shift. The blaze recedes. Sunday morning breaks. The fire burned in a horseshoe shape, passing Moorpark on either side. The zoo, the college, the surrounding neighborhoods are an oasis of green in a world of black. The only signs of life are a few old oak trees that somehow weathered the conflagration. Their time may still come. Brush fires spark here and there, their flames writhing in the hot wind. Throughout the day, everyone remains at the ready to evacuate. Crated animals are checked every half hour. They grow agitated in the closed, increasingly dirty quarters that are filling up with urine and feces. Buttercup, the badger, tries to dig her way out of her crate. Rowdy, the baby skunk, hides in a corner of his. The spider monkeys chatter nervously. A couple of students trail Bwana as he flaps around the teaching zoo, trying to recapture the turaco. Everyone's eyes smart and lungs strain from the smoke and ash. Everyone is exhausted. Some are punch-drunk. Dr. Peddie, who spent the night as well, tries to send some of them home to sleep, but no one will leave. Who can nap when the world may still go up in flames?

All the zoo's birds, reptiles, and some of the small mammals have been moved to another facility where they can breathe more easily. To the east, the fire licks at the Ronald Reagan Presidential Library. The highway that runs in front of the school remains closed. In the mid-afternoon, the ground shimmies underfoot. Wischhusen, taking a break in an air-conditioned classroom, looks up and says, "I think that was an earthquake." She's right.

During rounds, students notice that Tango, the cheery arctic fox, is breathing hard. However, he always pants; with his furred foot pads and dense coat, he's made for life on the permafrost, not in this parched landscape. He's also lethargic. Dr. Peddie is fetched. The fox is moved to an air-conditioned classroom. His body temperature is 108°. Tango must be chilled. The fox is attached to an IV, injected with steroids, and given a cool-water enema. His furry paws are dipped in ice water. His body temp dips to 101°. For a few hours the fox is stabilized and everyone breathes a sigh of relief. Suddenly, Tango pants his last pant. The teaching zoo has lost its first animal to the fire.

By Sunday night the fire recedes, so the animals are returned to their cages. Still, students and some staff stay overnight again just in case a passing ember should alight. Smoke still shrouds the zoo. Flickers spark on the horizon here and there. Over the next few days, the fires smolder, then die. The sky showers ash. It catches in everyone's hair. Students scrub the teaching zoo from top to bottom. Bwana, the turaco, flies around the zoo until he can't resist a handful of grapes held aloft. He returns to his cage. Kermit, a noble macaw, is found dead on the bottom of his cage, making him the second casualty. All the birds set loose from the aviary return, except for the pheasant.

It could have been so much worse. Most of the animals willingly were crated. Relatively few animals were hurt. Olive could have escaped but she didn't. Ignorance was bliss for the first years, who didn't fully realize the risks they took to crate certain of the various animals. If they had, they might not have evacuated as many. Everyone walks around with a renewed sense of life, that surge of optimism that follows a missed bullet. From the zoo, they look around at the scorched hills and feel incredibly lucky and profoundly aware of the rush of time. Life—these young students think for the first time—is short.

Maybe because of heightened emotions, after a few harmonious days the good feelings begin to evaporate. The second years, worried that all their training was undone by the evacuation, want to return to normal. That means no animal interaction for the first years. The first years, having risked their lives for some of the animals, can't let go. Haven't they earned the right to talk to the animals? Aren't they heroes? The second years think they are being dramatic.

After a couple of weeks of sniping, Dr. Peddie asks the college's psychologist to come up. Everyone gathers in Zoo 2. Even some staff members cry as they relive the fire. The psychologist explains post-traumatic stress syndrome. A second year accuses the first years of not respecting her class. A first year recounts how she feared for her life during the first evacuation. Someone says she would "die for her animals." Others raise their eyebrows and think to themselves, I would not. When it's over, some students feel better, some don't. Like a bad burn, hurt feelings can be a long time healing.

Elephants

Over the past three days April Matott has scooped elephant poop from dawn until dusk. The seven Asian elephants at Have Trunk Will Travel, a noted private facility, produce as much as fifteen wheelbarrow loads a day. It all must be cleared away with an elephant-sized pooper-scooper—a shovel. This is a novel experience for the second year. At EATM she became familiar with all kinds of dung, from the big snakes' squishy bowel movements, striated with undigested white rat fur, to the llama's beanlike droppings that bounce as they hit the ground. Elephant poop, however, has remained an unknown until now.

This is what the twenty-one-year-old has learned so far: pachyderms defecate nearly as often as sheep, but in lumps the size of melons or even loaves of bread. As guano goes, it's hardly smelly. The weedy potpourri of hay and grass smells something like chamomile tea spiked with nostril-dilating eucalyptus. What the dung lacks in stench it makes up for in heft. When the diminutive Matott lifts her shovel and stands at the ready behind a set of great gray haunches with a raised tail, she must flex her arms, because the soon-to-emerge dung will land with forceful kerplunks. If she doesn't hold on tight, the plummeting, steamy elephant missiles could knock the shovel right out of her hands.

This has not been the EATM student's only lesson during her brief internship here at the elephant ranch. In short order, she's learned that life with elephants is consuming, if not dangerous, increasingly controversial, and possibly heartbreaking. As EATM drums into the students nonstop, working with animals is a lifestyle, not a job. But to be an elephant trainer demands an unwavering devotion. The reward is a profound bond with the world's largest land mammal. The ranks of humans that can claim this privilege are small and growing smaller.

It is the second week of November, what they call project week around EATM, when the second years go on one-week internships. It is a chance, however brief, to leave the dreamlike but hermetic confines of EATM for a taste of life amid professional trainers and zookeepers. Once again, the second years leave the animals in the hands of first years who aren't too sorry to see them go. Relations have improved, but the tension between the two classes, like the smell of smoke that lingers around the teaching zoo, has yet to blow over.

Most of the second years haven't gone far for the week because California is rife with zoos, animal parks, aquariums, and studio companies. Becki Brunelli and Trevor Jahangard headed north to work with baboon trainer Kevin Keith in Vallejo. Mary VanHollebeke decamped for the Folsom City Zoo. Amy Mohelnitzky hooked up with a studio company in the Los Angeles area. Twenty-one-year-old Matott climbed into her '94 Plymouth, a high school graduation gift, and drove two hours southeast and checked into a cheap motel in Lake Elsinore, the town closest to Have Trunk.

Animal people typically break into groups roughly by species: marine mammals, big cats, primates, dogs, birds, or elephants. The last group is perhaps smallest, possibly because they have the fewest opportunities. Members of the American Zoo and Aquarium Association house some 305 elephants (160 Asian and 145 African). Private facilities such as Have Trunk care for another 200 to 300. They are tended by some 600 elephant trainers, according to a 1997 report by the U.S. Department of Labor.

Elephant trainers have one of the longest apprenticeships of any trainers, spending years learning how to control an animal twenty-five to fifty times their size. In addition to patience, they need guts. A slightly pissed off elephant can kill a keeper with a mere push or a kick. It's a perilous job, because elephant keepers and trainers traditionally have worked hands-on with the animals. At a zoo, you won't find keepers going in with the rhinos or the gorillas, but you may find them in with the elephants.

That age-old system, though, has begun to fade as more and more American zoos switch to what is called protected contact, which requires keepers to work with the elephants through the cage bars. With that change, the hands-on approach is not being passed along to a younger generation. Gary and Kari Johnson, owners of Have Trunk, worry that their way of working will soon be a lost art.

In fact, the whole profession may soon be on the endangered list. With the 1973 ban on importing Asian elephants plus the limited success of breeding in captivity, the number of elephants in North America is bound to dwindle. Elephants have also become the poster animal for People for the Ethical Treatment of Animals and its ilk. Circuses and zoos are pressured to retire their elephants to sanctuaries. Elephant trainers, good and bad alike, are smeared by animal rights protesters. Anyone with ideas about becoming an elephant trainer can't help but think twice. The road is long, uphill, and may lead nowhere.

The Johnsons at Have Trunk Will Travel take one EATM student for one week each year. This is one of the few shots an EATM student has at getting any elephant experience. The teaching zoo does not have a pachyderm for several reasons. Elephants are slow to make bonds and the teaching zoo's rotating student trainers would make these huge, hierarchical mammals hard to control and dangerous. Elephants are also expensive to keep, given that a typical one chews through about 200 pounds of hay a day.

Moreover, there aren't typically enough students interested in working with elephants to justify the cost, usually only one or two students per class, according to Dr. Peddie. Among the second years, they are Becki Brunelli and a young woman from Baltimore. Matott isn't one, so Dr. Peddie was surprised when she asked him to recommend her to Have Trunk.

Matott is a generalist. She hasn't cast her lot with one species. After EATM, she might like to work for an educational outreach facility, but she has an open mind about her future. All Matott knows is that she'd like to move back near her family in Albany, New York. Matott is small, her manner understated, so much so she can be easy to miss, especially among so many strong personalities. Her voice has a slight squeak to it. Her jackets and sweatshirts are always about two sizes too big and her straight dark-blond hair often a tad disheveled. She has a nose ring. She's paying her own way through school, working as a vet tech at an animal emergency clinic. She adopted a badly burned stray kitten that showed up at the clinic after the recent fire. The kitten joins Matott's other cat, as well as two rats and two hedgehogs.

In addition to training Wilhelmina, the hornbill, to step up onto a perch, Matott trained the teaching zoo's two guenons to get inside a crate. She also works with Abbey; Julietta, the emu; Sly, an opossum; one of the pigs; and Ghost, the clumsy bald eagle. Despite Matott's small size, whenever Ghost bates, lunging off her hand while wildly flapping his wings, she calmly swings the eagle in a neat circle back onto her gloved hand. Once the eagle gave her breast a good pinch with his hooked beak. Matott didn't flinch, just pursed her lips and blew a piece of hair that had fallen across her eyes out of the way.

Dr. Peddie is careful about who he sends to Have Trunk. First, the Johnsons are his close friends. Second, the Johnsons suffer no slouches, and by slouches they mean anyone who can't shovel elephant poop twelve hours a day. They are considered some of the top, if not *the* top, trainers in the country, Dr. Peddie says, and they run a facility that is tidier than a ritzy resort. Also, he wants to send someone humble enough to realize they aren't going to learn anything about training elephants. How could they in just a week? What they'll learn is how to sweep the barn, rake the crushed-granite yard, brush an elephant, and, of course, shovel poop. "It sounds like it's nothing but there's a lot to it," says Kari Johnson.

Have Trunk Will Travel is a ten-acre ranch tucked into the rolling hills of Perris, an impoverished rural community south of LA. You might never know the ranch was there, the way it's hidden behind a hill. As I pull up to the gate, I wonder if I'm lost even though I've followed Kari's directions, turning at a trailer with two satellite dishes aimed at the expansive blue sky. There is no sign of the small elephant herd. No trumpet calls, no flapping ears anywhere in sight. Through the entry gate, I can see the drive is lined with perfectly spaced palms and cypress trees cut into elongated barrel shapes. The gate rolls back. I slowly drive past a yard full of ponies. A large house to my left perches over grassy fields. Everything is spic and span and so still it seems deserted. Then I notice some movement. Two trunks wag in tandem. A pair of elephants stand side by side on a hillside. As I follow the drive it descends to a large lot that circles a barn, and I spot more trunks and flapping ears. Kari meets me at my car. She has a head of red hair. She comes across as ladylike yet earthy and strong. Like most animal trainers I meet, she has a physical assuredness about her. Her figure has spread some in middle age, but, as she says, she's "fat but fit."

The elephants have recently finished their morning regimen: a brisk, single file walk around the property, a snack of grass in a neighbor's field, a bath, and a training session. Kari points out who is who in the herd. That's Tess and Becky, the two best friends standing close together on the hillside. Tusko, their bull, is in an enclosure on one side of the barn. The four other females—Tai, Kitty, Dixie, and Rosie—keep him company this morning. No day is the same at Have Trunk Will Travel. The Johnsons are always changing the elephants' routine, putting them in different enclosures, and training them different behaviors to keep them from getting in a rut.

Kari leads me up to the office on the top floor of the barn with its black leather furniture, elephant knickknacks, and views of the compound. Out one window, I see Matott push a wheelbarrow of dung the color of golden fall leaves. Photos of the elephants at work cover the white office walls. In one, Tai holds Goldie Hawn aloft with her trunk. In another, Daryl Hannah lounges along Dixie's spine. An elephant balances on huge fake roller skates in an ad for Korean Airlines. "That was shot in the barn here," Kari says.

Kari goes off to collect Gary. Meanwhile, I chat with Joanne Smith, an EATM grad who has worked here for nine years—longer than anyone. Smith has on a full face of makeup and a baseball cap that reads I'm Having a Bad Hair Day. Smith was always nuts for elephants as far back as she can remember. EATM didn't really prepare her for her heart's desire, but, as she says, "You can't learn anything about training elephants from a book." When Smith graduated in 1989, she took a job with the elephants at the Racine Zoo in Wisconsin. "My mom was afraid, because you don't make much money in this business," she says. When the elephants were moved to warmer climes in Texas, Smith nearly followed but instead signed on with the Johnsons, because their "elephants can go everywhere. Most zoos' [elephants] don't do the amount of stuff they do." The herd travels around the country in a small fleet of semis. They play soccer and baseball with their trainers. They even paint. Smith prefers Dixie's nice long strokes to Kitty's slapdash style. The only trick, she says, is teaching them to keep dunking their brushes in the water jar.

"I just ended up doing exactly what I wanted," she says, and smiles. Not that it's been easy. Despite her long employ with the Johnsons, she still considers herself a trainer in training. Gary is also known for his exacting standards. Many other apprentice trainers have come and gone while Smith has been with the Johnsons. Her marriage caved under the long hours and demands. Ultimately, her husband didn't like the fact that the elephants came first. They have to, she says. Now she lives on the grounds, works every other weekend, every other night, and rarely socializes.

Smith runs off to the dentist. Out the window, I see Matott cross the lot with a rake in hand. Kari comes up the steps with Gary behind her. He has the bulky forearms of an old-time iron man. His thick hair sprouts in a widow's peak over his low broad forehead. His voice is deep and he speaks slowly. Gary got his first elephant in 1970 when he was sixteen. Back then, all you needed in order to get an elephant was the money. Gary didn't even have that. He traded a llama and two pygmy goats for his first elephant, Sammy, a male that was part of a small petting zoo at an Italian restaurant. Gary had a good idea what he was getting into. He'd worked at a petting zoo, sweeping poop at first but eventually graduating to running the elephant ride. When he was fourteen, he'd lied about his age and spent a summer with Ringling Bros. working with an elephant. Eventually he was kicked out of school for playing hooky so he could spend an afternoon watching elephants at a shopping center. Still, he admits that buying his own was "a weirdo thing to do."

Gary kept Sammy on his parents' twenty-acre ranch in Anaheim. He housed him in his family's oversized garage and trained him mostly by trial and error. Then the elephant reached sexual maturity and got too aggressive. Gary sent Sammy to a park in Mexico. In 1978, he got Tai, the herd's thirty-five-year-old matriarch and star performer. As Kari likes to say, "Gary got Tai before I got Gary." After all these years, elephants still have their hold on him. He'll forget to come in for meals, Kari says. "He's hard to manage. You can't make him not do it."

Kari fell into elephant training. It was the family business. Her stepfather was Robert "Smokey" Jones, a world famous elephant trainer. He trained circus and zoo elephants around the country. With his partner, he trained an elephant to water ski. Traditionally, elephants have been trained through dominance and often heavy use of the bull hook. Jones used a more enlightened approach that relied on cooperation and consistency. He studied the elephants and got to know their habits, the way their minds worked, just as Gary and Kari have.

Kari wasn't all that interested in elephant training but she began working with them when she was thirteen or fourteen. She was fifteen when Gary began hanging around her stepdad. She thought he was a hick. Ten or so years later, Kari finally noticed how cute he was. By then, Gary had three elephants. Unlike most women, Kari didn't mind the package deal. The couple was so busy living and breathing elephants they never really dated. "We meant to," Kari says. "We just moved in together." While they were at a convention in Las Vegas, Gary asked Kari, "Do you want to go ahead and get married?" They squeezed the nuptials into a busy day, went back to their room, and toasted each other with Diet Coke poured into two cups from the hotel bathroom. Elephants encircle their wedding bands.

Keeping your own elephant is expensive. You need not only mountains of hay but also a place big enough to keep the animal. However, an elephant doesn't really leave time for having a job. You need to pay for the elephant with the elephant. So an elephant naturally becomes your job and your life, as it has with the Johnsons.

Movie work pays the best. Gary and Tai have made a long list of films, including *George of the Jungle* and *The Jungle Book*. Film jobs, unfortunately, are sporadic. So the couple take their elephants to fairs, zoos, and even pumpkin patches, where people clamber on without realizing they are on the back of a Hollywood star. Lately, they've been hiring out Tai for traditional Indian weddings, in which the groom, dressed in white from head to toe, sits in a special saddle and rides the animal to the ceremony.

The Johnsons are obviously happy with their unusual life. There are just two clouds on their otherwise clear horizon. The first is obvious: the animal rights movement. They picket the elephant rides the Johnsons offer at the Santa Ana Zoo. They have tried to ban exotics in communities around California, including the Johnsons' own Riverside County. If the elephants can't work, the Johnsons would not be able to afford to keep them. It hasn't come to that, but they worry. All the fighting and the constant testifying take them away from their work. When the Riverside ban was proposed, the Johnsons had to let their business go to fight the ordinance. Kari has testified at city councils, legislative committees, and even before the U.S. Congress. The Johnsons can't help feeling defensive, so much so that when I ask them if they rescued their elephants, Kari bristles and says, "We got them as they became available." She admits that they have paid more than they should have for some of their elephants "to get them out of a situation," but she won't use the word *rescue*. It's "too self important," she says. She won't use the enemy's vocabulary.

If Kari seems bitter, it's worth noting that the animal rights people kicked her when she was down, which brings us to the other dark cloud. This one, which has hung so low over the ranch for nearly five years, may soon lift.

I follow Kari outside for a tour of the compound. We start with Tusko, the 12,000-pound bull with a billboard-sized forehead. He is twice the size of any of the females. His dun-colored skin has an apricot sheen from all the dust he's sprayed on himself. He is the only elephant they cannot go in the enclosure with. Male elephants are much more dangerous because they are so big and aggressive, especially when they go into musth.

Tusko is part of the Johnsons' plan for the future—not just their own future but the future of Asian elephants. The Johnsons' income not only keeps the elephants in hay, the ranch in decomposed granite, and the Johnsons in their house but also funds the couple's breeding program. They are part of a countrywide effort to breed Asian elephants in captivity. As wild Asian elephant populations dwindle, people like the Johnsons believe that breeding captive animals will save the species. They are members of the AZA's Species Survival Plan Program. They are also one of the few private facilities that is AZA-accredited.

To this end, Tusko, the behemoth, was added to their herd. However, their jumbo-sized lothario was a disappointment at first. He didn't seem to know where to put his huge winky-tink, as Kari calls it, so they sent him off to a breeding facility for some much needed practice. After about a year, Tusko knocked up a female there. The Johnsons sent for their stud.

Kari walks me down a line of shiny gray trucks, including two semis neatly parked at the far end of the compound. They were especially designed by Gary for transporting the girls. They have air conditioning. One has a screen so the elephants can see out, but people can't see in. There are other clever inventions at the ranch, such as a Dumpster set in the ground and landscaped right up to the edges. You can't see it's full of poop until you're standing nearly on top of it.

As we round one end of the barn, we see the end of a trunk snaking over a fence delicately reaching for a flowering bush. "I see you miss Dixie," Kari calls. Dixie is the oldest at nearly forty, but she's not nearly as big as Tai, who has about a thousand pounds on her. Dixie is a talker. She even moos like a cow. Dixie, Kitty, Tai, and Rosie have moved over to the enclosure next door to Tusko. Here, they coat themselves with dust, elephant sunblock basically. They snack on the branches of the eucalyptus, cottonwood, and pine willow trees. They play, running and throwing sticks at each other, and they nap. As Tai settles down for a late-morning snooze, going down on one knee, then rolling slowly onto her side, the other three elephants gather round like ladies in waiting.

In the hillside enclosure, Tess and Becky continue to move in tandem like Siamese twins. These best friends may have a secret in common: both mated with Tusko this fall. Both might be pregnant. It's about ten weeks before you can know for sure. Tusko will soon get a third time at bat. Rosie is due to cycle soon. If all goes well, there may be three babies on the way. These wouldn't be the first babies, and this is what makes Kari's eyes go bright with that sharp brightness that precedes tears, though she can now talk about Amos and Annie without crying.

Five years ago, two baby elephants frolicked on this ranch—two wonderful, goofy baby elephants. Tess gave birth to 300-pound Annie with Dr. Peddie's help on July 30, 1998. She arrived feet first and got up right away. Annie looked like her mom. Her ears came to a pretty point and her smooth skin was dark. A month later, Becky had Amos after a long delivery. He was very hairy like his mom. Amos didn't have Annie's pizzazz but was lovable all the same. Everyone fell hard for these two babies, who splashed in their Barbie swimming pool and chased each other all day long. Like a proud mom, Kari took their photos to the bank and the tellers passed them around. Joanne Smith secretly taught Amos to paint and gave the Johnsons his masterpiece for a Christmas present. The Johnsons had a TV monitor in their room so they could keep an eye on the babies at night. They would stay up late watching the twosome wrestle and play. Gary called it E-TV.

On an early spring morning in 1999, Annie suddenly became ill. Dr. Peddie was out of town, but two top elephant vets happened to be in the area. They were called to the ranch. Annie's lungs were filling with fluid. The eight-month-old elephant was given intravenous antibiotics but she got sicker and sicker. At the end of the day, as Kari conferred with Dr. Peddie on the phone, Annie's heart stopped.

Kari grieved like a bereaved mother, long and hard. She wished she could be like Tess, who trumpeted and paced inconsolably for days and then got over it. An animal rights activist accused the Johnsons of having separated Annie from her mother and thus prompting her death. In fact, mother and daughter had never been apart. The autopsy showed nothing conclusive. All the Johnsons knew was that they had two baby elephants and then they had one.

Then they had none.

That summer, Amos succumbed to a twisted intestine on Gary's birthday, August 8. Joanne Smith says it was the worst thing she ever went through. Dr. Peddie says he's tried to block the deaths out of his memory. "Those babies were the best thing that ever happened to us," Kari says. "Losing them was the worst thing that ever happened."

* * *

Matott has been cleaning alongside David Smith, the Johnsons' newest employee, since 6:30 A.M. If either is tired, it doesn't show. Smith, who is not related to Joanne Smith, graduated from EATM just last spring. He has a lithe build, a thin face with a small chin, and close-cropped hair. He is friendly but serious, weighing his words as he speaks. He lives in a trailer on the property and gets to see his girlfriend only once a week on his day off. He works twelve-hour days at a minimum and doesn't make much money. He has a master's degree.

Smith was an environmental technician in the Los Angeles area for ten years. He began to wonder if he was making a difference, if the long days were worth it. His volunteer work at an animal refuge made him decide to go to EATM at thirty-four. Smith did a one-week internship with Have Trunk last fall. He didn't have his heart set on working with elephants, but the Johnsons' conservation work appealed to the environmentalist in him. The Johnsons hired him when he graduated. His primary duty now is to clean the ranch, but eventually the Johnsons hope to make a free-contact elephant trainer out of him and thus pass on their method to a younger generation. "It is kind of amazing, what an opportunity it is," Dave says. "There's nothing like this situation in the country."

This apprenticeship is slow going; it not only takes time to bond with an elephant (as much as two years) but also to master the fine nuances of the profession. Over the years, the Johnsons have started to teach a half dozen or so would-be trainers, only to have them give up midstream. By taking Smith on, they are gambling a good chunk of time. It will be a couple of years before they'll know whether Smith can develop the instinct to be two to three steps ahead of an elephant. If he doesn't, "he'll still be a good guy," Kari says. "He just won't be a great elephant guy." When I ask Smith how he feels about the long road ahead, he says, "Best not to entertain those thoughts."

He took some baby steps this summer, and picked up pointers, such as elephants respond best to quiet, calm commands. He got acquainted with the herd and they with him. Mostly, though, he helped getting the elephants back and forth from various fairs and festivals. Since September, he's been working on training each day with Becky or Rosie. He finds there's such a big gap in knowledge between him and the other trainers, they can't always explain to him what he's doing wrong. He gets discouraged. "Handling elephants is supremely more difficult than it looks," he says.

Not long after lunch, he and Matott briefly set aside their shovels and wheelbarrows. Becky, Smith's practice elephant, is collected from the hillside enclosure. She is sociable and a good sport. She's young at only twenty-five. She will still add another thousand pounds over the next ten years. She's the hairiest of the herd, with a fuzzy halo of wiry black hair on her brow and shoulders. She has ugly rough calluses on her temples that resemble patches of gravel, but as Kari says, "She's so cute it doesn't matter." At this point, Becky knows far more about elephant training than Dave does.

Kari has Dave start by walking Becky in a circle. They stroll around quickly, Becky keeping abreast of Dave's shoulder. He carries a bull hook in his hand, but like the Johnsons, rarely uses it and then only as a guide. The Johnsons, like Kari's stepdad, rely on rewards and consistency to train their elephants. They don't call it operant conditioning, but that is essentially what they are using. Kari, Matott, and I stand in the middle of the circle, turning and craning our necks as Smith and Becky circle us. The point is to keep Becky by his side, not to let her lag behind or surge ahead. This is what Dave has to learn first—how to lead Becky. "If you can't do that, you can't do anything," Kari says.

Kari tells Dave to stop and have Becky swing around. "Get in line," he commands quietly. Becky stops and turns to face him.

"Steady," he says. Becky's body should be at a neat ninety-degree angle to Dave. If she's not, that's her way of testing the novice trainer. The bulk of an elephant is hard to size up. Not hearing any encouragement from Kari, Dave turns to look at her. "Where's the front?" Kari asks. Dave looks as if he isn't quite sure. He good-naturedly walks around Becky and keeps asking her to move a bit here, a bit there. It's like trying to get an SUV perfectly straight in a parking space. Becky moves a bit forward, and her rear end goes out of whack. "She's past center," Kari says. "Do you see the front?"

If Kari has a note of impatience in her voice, Smith seems unfazed. He finally gets Becky into a correct position and then says, "Trunk." Becky raises her trunk high. "Foot," Smith says and Becky, as crows squawk nearby, daintily lifts one rounded front foot so that you can see its sole is pink. Then he walks a circle around her. "Can you tell April why you are doing that?" Kari asks.

"I don't want her to think that she can move when I move," he says.

Off they go again, man and elephant padding around and around in the afternoon sun. Dave stops Becky whenever Kari tells him to and goes through the same drill over and over. "Get in line," he says, and Becky swings her bulk around. "Steady," he says, then straightens her out. Dave has her raise her trunk and then a foot, and they return to their hypnotizing circles. After several attempts, Smith nails it on his first try: Becky stops on a dime and swings into place with the precision of a Marine.

"She's very straight. Good," Kari says.

After a few more spins around, Kari stops them so Becky can go to the bathroom. Matott dashes behind Becky with a shovel while Kari commands, "Go potty." Becky rumbles, squeaks like air being let out of a balloon, and then roars. One loaf pops out and hits April's waiting shovel. The afternoon wanes and there is yet more cleaning, more poop to be tidied up. That's enough practice for one day. Dave strides through the gate into the hillside enclosure with Becky right by his side. "Now, Dave, you know better than that," Kari scolds. An elephant as placid as Becky could inadvertently smash him against the gate. Kari's reprimand breaks the quiet magic of the past half hour. It is a reminder that just one little mistake could have dire consequences. This is one of the tricks of elephant training—never forgetting for one second that these intelligent, personable beasts could hurt or even kill you.

Dave and Matott rush over to feed the ponies, Gary's "hobby," then back to ready the elephants for the night. In the fading light, they scoop Tusko's enclosure while the bull devours hay and loudly slurps water. The female elephants dip down on one knee and roll on their sides for a brushing. Then they rise and clasp one another's tails with their trunks and step lightly out of their enclosure. They stop and swing to face Kari, who commands, "Get busy." Kari walks between the elephants, chirping, and tickles their bellies with her fingers. "Oooo, ooo," she coos. There is much rumbling and straining. Matott is back in position behind the elephants, trying to gauge who's about to let loose. As she gives up on one raised tail to step to another, a steamy lump tumbles out. She swings her shovel back but it's too late. "I missed it," she wails.

"Tails," Kari says. The four females take one another's tails again and walk single file through the deepening night and into the barn. Soon the roomy barn fills up with their swaying bulk. Most of the night, Kari says, they will eat. They won't sleep until near dawn. The girls line up in specific order, so best friends are next to each other. From left to right, it goes Becky, Tess, then the mischievous Kitty, the matriarch Tai, chatty Dixie, and little Rosie. One foot in front and one foot in back is chained to keep them in this order. This makes for a peaceful night and keeps them from eating each other's food—well, almost.

The noshing has already begun. Becky eats from her pile of hay and helps herself to Tess and Kitty's on either side. Rosie pushes her pile out of reach of the herd. Tai doesn't seem to care who snacks from her stack of hay. Though they all defecated outside, they are already at it again in the barn. Matott and Smith push a wheelbarrow down the line of large bottoms. Then, though their long day is officially done, the two humans slouch on hay bales and watch the elephants dine. Elephants may move with an uncanny silence given their size but they are noisy eaters. They smack their great lips and grind their teeth. They breathe heavily. They play with their hay using their trunks. They chew and chew in great circular motions. It is mesmerizing. Eventually, Matott and Smith rouse themselves and reluctantly leave. Matott has a test to study for. Both have another early morning ahead.

On the last day of her week, Matott is rewarded. The Johnsons give her a T-shirt that Rosie and Tess painted. She climbs atop Tai for a ride. Matott lies on the ground as the elephant leans over, gently takes the young woman's leg in her mouth, and then lifts her. Though Tai's teeth pinch a little, Matott smiles for the camera.

It's been a good week, Matott thinks. She's learned a lot, including that elephant training is not for her, at least not right now. She noticed that everyone who works for Have Trunk Will Travel lives there and that they rarely leave the compound without the elephants. She wants to work with a range of animals. She also wants time for her own animals, for friends, and for other experiences. She's not quite ready to devote herself to one all-consuming species, even if it's the beautiful, mighty, and intelligent elephant. She's young. Matott's not ready to give her life away, at least not yet.

November

Back at EATM, Austin and Alamo, the once adorable prairie dogs that arrived earlier this fall, have gone into rut. They now prefer biting to cuddling. Austin nailed Castaneda, nicking her on the leg, then digging his little incisors into her thumb. It was her first bite. "It seems silly to be bitten by a herbivore," she tells me.

Leftover Halloween pumpkins are stacked outside of Nutrition. The big cats will get to swat them around like gazelle heads. The November chill has chased the flies out of Hoofstock. Tremor and Little Joe, the tortoises, have been moved into a tub in the Reptile Room. Legend's fur thickens. So does Nick's. The miniature horse remains plenty plump no matter what Jena Anderson does. She tried putting the llama's food on an upturned garbage can lid outside the corral beyond Nick's reach. The lid, though, fills with water when it rains. She'll have to think of something else.

With the second years gone on projects, the first years work at triple time. Terri Fidone, the former Vegas cocktail waitress, babysits

Abbey while Mohelnitzky is away. It's like caring for a heartsick lover. Abbey shreds her blankets. She throws up more than usual. She refuses breakfast though Fidone douses her food with scrambled eggs and barbecue sauce, as Mohelnitzky instructed. At least the pooch has moved to better quarters, the arctic fox's old digs at the front of Quarantine. Now Abbey can see the goings on in Primate Gardens, such as the binturong, Chance, being fed or students training the lemurs, Obi-wan and Jenga.

Without the second years the zoo feels slightly deserted. On a Wednesday afternoon, I find only a few students trundling about in their wellies, eyes glued to flash cards in their hands. Looking south over the green athletic fields, the banks of the foothills are an ugly black in every direction. The faintest scent of wood smoke wafts over the teaching zoo. In Primate Gardens, Zulu chomps a head of romaine lettuce while watching TV. Someone has plugged the small black-and-white set into a heavy orange extension cord and set it on a chair outside his cage. The sound is turned off. The baboon sits on his puffed pink bottom, motionless except for absentmindedly scratching himself here and there, while he watches a car commercial.

I run into Castaneda, who's on her way to feed the cougar brothers a slippery mix of chicken necks and ground mystery meat shaped into balls. As we walk up the back road, we look north toward the tarnished landscape beyond the zoo's lush green. "How do you like the Happy Face?" Linda asks me, pointing out the maniacally smiling cartoon someone has etched into a charred hillside. The flames have long since receded, but the fire burns on. Mary VanHollebeke has had to start over from the beginning with Starsky's training. Ash still accumulates in the animals' cages. Mara Rodriguez's car only recently ceased smelling of smoke. Work has begun on an emergency plan for a fire. Goblin, the baboon, is being crate trained.

Inside Big Carns, Castaneda and another first year go to opposite sides of the cage and each calls over one of the cougar brothers. Castaneda gets Sage to climb onto a high shelf in his cage and then starts pushing meatballs through the bars into the cat's waiting mouth with the palms of her hands. Castaneda gingerly offers a chicken neck through the bars so he has to tip his head up to munch it. He crunches the bones loudly. When you hear that, Castaneda says, you can't help wondering if they snagged your finger. Cage bars are no guarantee against getting hurt. Samburu recently hooked a second year's finger, splitting it right up the meaty tip, while she was feeding him through the bars.

"I love you all," Castaneda says to her fingers. "If I lose one, I'll miss you. Nothing personal." The cougars purr and make happy kitty sounds. Castaneda chatters in response. "Isn't that yummy?" Savuti, the hyena, stands at attention one cage over, watching the feast. Susan Patch pauses along the front road to watch.

"Why do you hold the food high up like that?" Patch asks Castaneda.

"Holly said to, because if he eats too fast he'll throw up."

Patch looks dubious. She often does. Patch's gotten a bit of a reputation as a know-it-all, but she *does* have a degree in animal science. Patch's skin, like a lot of EATM students, has broken out. Unlike a lot of EATM students, she has lost weight. A former competitive rower, Patch thinks it's because she's losing muscle mass. She'd be the only one. Biceps and abs are generally on the increase. So are bellies and bottoms. Midriffs lap over khakis. Upper arms swell and jiggle. It seems impossible that you could gain weight here, but then you see the grab bags of candy passed around classrooms, the cartons of steamy fast food imported by the hour, and the boxes of slicked doughnuts at every kind of meeting.

The strain of this first semester takes its toll. EATM exacerbates any problems or flaws the students have. Larissa Comb is heartbroken over a relationship that ended. At EATM, where there's no time to date, her loneliness deepens. Another first year counts her pennies and pays for her groceries with food stamps. This is Chandra Cohn, the lean, long-faced first year who bawled loudly over pulling her pigeon. Just before orientation her husband asked her for a separation. Now she's scrambling to pay the rent on the room that she, her daughter, a ferret, two rats, and two hermit crabs share in a houseful of students. Castaneda's rheumatoid arthritis has flared, not to mention her allergies. When she worked in the Rat Room, her eyes became swollen even though she wore a face mask. She got a sinus infection. Her perfectionist ways also get the better of her. She copies her classroom notes, writing them out in her neat print so they are "pretty." EATM doesn't leave time for getting things just right, but Castaneda can't stop herself.

One of the rules of training is to not let your animal get too frustrated, but EATM does not always practice with its own students what it preaches. First years agree they would do better if they could just talk to the animals, touch them. They are, after all, animal people. Still, nobody drops out as students did in the past.

Sage passes on the last chicken neck Castaneda offers. "That's funny," she says. Castaneda holds her hand up against the cougar's cage. As Sage licks the meaty gunk off her palm we hear someone call, "I love you Schmoo." Schmoo—not one of Castaneda's favorite animals these days! While the cougar happily takes nearly every morsel Castaneda tenders, the sea lion will not. With the second years gone, Schmoo has resumed her hunger strike. I follow Castaneda out of Big Carns and down the back road toward Schmoo. As we walk, we can hear the sea lion's low, guttural rumble, almost like an old man clearing his throat. As a tall, redheaded first year passes us, she blurts out proudly, "My monkey caught a bird."

We meet up with the other temporary Schmoo girls and Gary Wilson, who's come to see if he can tempt the sea lion to eat. Schmoo's pool gurgles. Sirocco and Kaleb, the camels, turn their large snouts our way. Wilson has an idea: forget putting the pills in the fish. Wilson calls Schmoo over and opens the gate. She waddles out, her dark eyes shining, her extra-long whiskers glistening in the sun. Wilson rattles off commands, to which Schmoo energetically responds, throwing her tail up, waving a flipper, and rolling over on the hard-packed earth. "Open!" Wilson says. The sea lion opens her mouth, exposing two banks of black teeth and a bubblegum-pink tongue. One of the temporary Schmoo girls, a blonde with a high voice, quickly tosses a handful of pills in. Mission accomplished.

Wilson keeps working Schmoo, commanding her to turn left, to circle, wiggle her whiskers, and roll over. He rewards her with fish and squid, which she gobbles down. By now, her dark stomach is pebbled with light specks of dirt. "Touch!" Wilson says and rubs her belly. Schmoo opens her mouth as if it tickles, then downs a glob of squid Wilson tosses her. The temporary Schmoo girls stand in the shade watching, not smiling.

"She wouldn't even take fish from us," Castaneda harrumphs.

"She's acted like she hated us," the blonde says.

"She snubs us," Castaneda says.

"I swear she rolls her eyes at us," the blonde counters.

"Look how happy she is with him," Castaneda says.

Suddenly Schmoo lunges at Wilson with her teeth bared. He jumps back. The Schmoo girls tense. "Relax," he says. The sea lion closes her mouth and loosens her posture. The temporary Schmoo girls, though, remain rigid.

Brenda Woodhouse is not a morning person by nature. Nevertheless, five days a week she rises at 4:30 A.M., when night still has its black hold on the sky, and drives fifty-two miles from her home in Acton to EATM. She's due at the school before 6:30 A.M., when she starts calling the roll in her clear, strong voice and marking down who's late. She says she never gets enough sleep, but it doesn't show. At this early hour, when the students hunch over their desks and recede into their sweatshirts, Woodhouse, who naturally walks fast, talks fast, and thinks fast, is on full alert.

This morning, like all mornings, she's here, there, and everywhere at the zoo, supervising inexperienced students as best she can, answering questions as she goes. I have to trot to keep up with her as she strides down the back road. As she rounds Nutrition, a first year dashes over to her, something grasped in her hand. She opens her fingers. It's a baby rat, a pinkie, with its shoulder bone sticking out. "We're out of gas," the student says. Woodhouse suggests crushing it quickly with a rock.

Woodhouse is heavy set and strong. Her blond hair falls to her shoulders. She has wide-set eyes and a rounded nose. She moves her hands a lot when she talks. She's magnetic. Growing up, she was the link between her deaf parents and the world of sound. Her parents signed at home but only lip-read in public; signing had a stigma then. Both could speak, Woodhouse says, in uninflected monotones. Still, they leaned on their only child. It may have been too much responsibility for a child, but Woodhouse's unusual upbringing prepared her for communicating in a nonverbal world. "I grew up reading body language," she says.

She thought she'd work with deaf people and studied psychology in college. Then she found out about EATM and changed course. She would be a different kind of interpreter, a link between the animal world and our own. It was the Briz years. Woodhouse didn't think she'd made much of an impression on him because he always called her Louise. She left the program two months early without graduating to take her dream job, working with marine mammals at Chicago's Brookfield Zoo. She loved it there among the zoo's five dolphins, three California sea lions, one walrus, and one harbor seal. It was a union job that paid well. She had not worked with EATM's sole marine mammal, Schmoo 1, a more affable sea lion than the current Schmoo, but Woodhouse learned on the job. There were only two things she didn't like: jumping in and out of the chilly dolphin pools in the winter and handling endless buckets of fish. A visitor once asked her how she got dates, smelling like fish. In fact, she was married to her high school sweetheart. She left her job after she had the first of her two kids.

Her family moved back west. Her husband is a high school guidance counselor. Woodhouse took a part-time job with the bird show at the Los Angeles Zoo. She also trained the drills, large, tailess monkeys related to mandrills, for artificial insemination. Then in 1994, she joined the EATM faculty. She's an upbeat, gregarious teacher. Animals still have a hold on her, especially anything with hooves, but now her focus is on her own species.

Just a few steps farther past Nutrition, another student approaches Woodhouse. "Can Zulu have this?" she says, holding out a toy green frog with stuffing coming out. "No," Woodhouse says. The student disappears into the barn and returns with a stuffed, floppy-eared, purple bunny with its button eyes removed. "Yes," Woodhouse rules.

We all walk over to his cage. Zulu grabs a rubber wellie and slowly climbs onto one of his shelves. With his elaborate coloring, Zulu is at times regal, at times clownish. A stripe of cherry red runs down the middle of his face over his nose and upper lip. To either side of the stripe, his cheeks are a faded blue. His beard is a lemony blond, his chest hair a creamy white, and his balls an impossible violet.

His pace is unhurried. He's rather aloof compared to the busy, chattering spider monkeys next door. He spends much of his day quietly watching the students scurry about. However, should you approach his cage, his lips will spread in that exaggerated clownish grin with the yellow canines overlapping his bottom lip. It's quite a sight. The rule with primates is that a smile suggests fear.

With Zulu, though, a smile is what it looks like, a howdy-do. He is nearly ten. He's been living at the teaching zoo for five years. He is owned by a professional trainer, Monty Cox, and his wife, Anna, who bought him when he was seven weeks old. Anna raised him at home, even teaching him to sit at her breakfast bar and eat with a spoon. By the time his canines came in, Zulu was far less cute and far more dangerous. Also, the Coxes did not have a permit to keep him in their home. EATM took the mandrill in.

At the zoo, Zulu whiles away his day with his collection of all things rubber: boots, balls, even a basketball. He loves bubble gum. The colors on his face light up when he's handed a new rubber toy. He'll drool. He often masturbates, tucking a wellie between his legs, rubbing it just so. This embarrasses, if not revolts, some of the students. They stop by his cage to say hello, only to notice the jerking of his hips; and with a surprised "Oh!" they turn on their heels. Woodhouse says all the male primates at the teaching zoo masturbate a good deal, but she's never seen one use a tool before.

Sunni Robertson, a cheerful first year with olive skin, is taking care of his highness this week. Robertson is trying him on for size. She might ask to be his trainer next year. He is quite an undertaking, as his current second year trainers will tell you. First, there is the pinching. Zulu can reach his hairy arm through his cage. His student trainers often sit within reach of the mandrill to groom him or just to keep him company. Should they annoy him, say by moving too quickly while grooming, Zulu will reach through the cage bars and squeeze their arms, pressing his black nails into their skin with relish. Moreover, if a student trainer ticks him off in general, maybe by talking to a friend near his cage when he wants her full attention, she is instructed to hold her arms up to the cage and let him pinch them, which he does. Otherwise, the thinking goes, he will stay mad. Having vented his rage, Zulu gets over it. The student trainer, having been grabbed, gets a bruise. He's like an abusive boyfriend. My guess is that no man would agree to this, but all Zulu's female trainers submit to his punishing grip.

Zulu offers the most painful lesson in how a trainer must consider an animal's natural history. In the wild, as Cindy Wilson explains it, Zulu would live with a harem of female mandrills. They would be with him all the time. Here at EATM, the student trainers substitute for his harem, but they disappear for long stretches every night, so Zulu, like an insecure king, has to keep reasserting his male dominance over his female subjects.

Certainly his headaches don't improve his mood. Whenever he gets a bad one, Zulu pulls on an ear. Bright lights will bother his eyes. He can even seem temporarily blind. No one knows exactly what to do for him. He's never had an MRI, Dr. Peddie says, because Anna is afraid to have him anesthetized. Zulu is a package deal. Anna visits him regularly and still has a fair amount to say about his care. Many students don't want to deal with someone who is not on the school's staff. Also, many of the students don't approve of Anna because she feeds Zulu burritos, pumpkin pies, and other junk food fit for a teenager. The mandrill has quite a beer gut, and the students blame Anna for it though she comes only every couple of weeks or so.

That's a lot to consider, but this week has gone well so far, Robertson says. Zulu hasn't scared her off, but she's not sure she's up to the pinching. While we stand by his cage talking, the king sits above us on a shelf in his cage, holding his scepter in one hand—a wellie.

At 8:30 A.M., the appointed time for Schmoo's morning feeding, the temporary Schmoo girl with blond hair and a high voice arrives at the sea lion's cage with a slick pile of freshly thawed fish and squid and a handful of pills. When she calls Schmoo over, she immediately obliges. "Open!" the student chirps and to our mutual shock, Schmoo does. The first year fumbles the pills and before she can toss them into Schmoo's mouth, the sea lion closes her trap. The first year groans. "Open! Open! Open!" she chirps, but nothing doing. Schmoo turns, waddles back to her pool, dives in one side and out another, slides into her favorite corner, and lifts her whiskered nose to the sun.

"Schmoo, you big brat," the first year says. "You don't want to listen to me. You're being silly." She calls Schmoo over again. The sea lion indulges her and scoots back over. "Open!" she tries again. A moment or two passes. We stare at her. Will she, won't she? The sea lion opens wide. The first year tosses in pill after pill, jumps, and pumps her fists in the air.

"I love you Schmoo," she squeals.

Schmoo isn't quite done being finicky, however. She won't eat any fish or squid. That the diva deigned to take the pills is more than enough. In fact, the student reports, it feels like Christmas.

Later, when I report to Castaneda that Schmoo finally took her pills, she looks equally relieved and irked. She would have liked to be the one who got the sea lion to take her meds. She has straight A's and a bachelor's from a respected college. She managed teenage punks in an inner-city school. She lived in an African jungle for weeks. Still, an elderly, nearly blind sea lion has gotten the better of this perfectionist.

The next day, a Saturday, the cleaning starts at a leisurely 8 A.M. There are no classes and the teaching zoo doesn't open to the public until 11 A.M., so the pace slows just a hair. In Hoofstock, two easygoing EATM students fill a wheelbarrow with poop from the sheep, the pigs, the deer, and the camels. Another first year stops by to pluck a log or two from the wheelbarrow for Savuti the hyena; he likes to roll in it. Sirocco and Kaleb watch the cleaning proceed with their heavily lashed eyes. Schmoo sits in her corner, nose turned skyward.

Now that the flies have waned, this is one of my favorite spots in the zoo. With pens rather than cages, it's airy and sunny. The sounds are soothing, from the sheep's rhythmic baaing to the water lapping in Schmoo's pool. The animals are mostly familiar, knowable. Though you have to be careful that you don't get too relaxed. As I'm talking to Adam Hyde, a towering, soft-fleshed eighteen-year-old who wears clothes two sizes too big, Sirocco, the white camel (the one with a bad reputation), leans over and presses his nose against Adam's arm. Hyde jumps and nearly trips over a salt lick. The other first year and I freeze. "That was enough to scare the hell out of me," Hyde says. He turns to Sirocco. "Putz," he says, recovering his teenage swagger. We laugh.

Castaneda, with her sure stride, comes through the gate, clutching Schmoo's plastic pill holder. From across Hoofstock, I watch as she calls Schmoo over. The sea lion, who always moves faster than you expect, zips right over. "Open!" Castaneda commands. Schmoo obliges. Castaneda tosses the pills in matter-of-factly. No fist pumping, no jumping up and down or squealing.

"It's nice not to be responsible for killing the sea lion," Castaneda says and turns on her heel and walks off. It's too little too late. This star student will pass on the star animal. She will not ask for Schmoo.

Baboon Here!

One December morning I arrive at the teaching zoo to find Rosie, the olive baboon, and Trevor Jahangard sitting side by side in the shade. They have settled on a bench not far from the front gate overlooking the Aviary and Clarence. Below, a bunny hops through the huge tortoise's lettuce-strewn enclosure. Clarence doesn't notice. He's trained his beady black eyes on the unlikely pair—the man and the baboon.

Rosie and Trevor resemble two old friends quietly enjoying each other's company on a sunny morning. It is brisk, but Jahangard has on a muscle shirt and shorts, which he seems to favor no matter the temperature. He clutches one end of Rosie's leash, which hooks to a belt low on the monkey's hips. Her legs stick out straight from the bench. Her expression is serene, though her copper-colored eyes look a tad serious. One black hand is tucked under a leg. Chris Jenkins leans on a wooden cane nearby. When I approach the threesome, he calls, "Baboon here!" I stop to chat, though I know I'm intruding on a male enclave of a sort.

That there are so many female students at EATM is generally a blessing for the few males in the program but a bit of a mixed one. Alum Mark Forbes says he loved being with so many girls and dated three of them. Once when Adam Hyde complained he needed a break from the "estrogen fest," a female student shot back: "Pu-leeze, you love it. You've never been around so many gorgeous women in your life." If the men are single, the school is a testosterone-laden male's dream come true. Numbers and time are on their side. There are currently 6 male students and 89 females, roughly a 1 to 14 ratio. EATM leaves the women little time to hunt for men elsewhere. As Rodriguez tells me, even the biggest dorks get laid at EATM. Of the six males in the two classes, the oldest is married, and Jenkins lives with his longtime, non-EATM girlfriend. That improves the ratio to 1 to 22. Jahangard kind of dates one of his fellow second years, a sharp, striking woman nearly ten years his senior. One of the first-year males is in an on-again, off-again romance with a classmate. Take them out of the running, and now the ratio is 1 to 44. Though Adam Hyde is just eighteen, chunky, and still has a high school smart-alecky bravado that doesn't play well with women over sixteen, the odds are on his side.

On the downside, the preponderance of women makes for all the presents and hugging. Even alpha females, who brave cougars and regularly hoist fifty-pound bags of feed, can get to squealing and trilling, not to mention gossiping. Jahangard tells me the women are more quarrelsome, that they have to talk about everything, and he doesn't need to. "I'm like, 'It's okay. It doesn't matter,' " he says.

When Jahangard and Jenkins crave some quiet male bonding, they can briefly leave the girliness behind and get Rosie out. Rosie is twenty. She came to EATM at four months. Her deep olive green coat is speckled with white. She weighs maybe thirty pounds. This is the first of Rosie's two daily walks. Jenkins, as backup, comes along each time. The threesome's walk can be an efficient twenty-minute turn around the compound or a lounge in the sun on the bleachers in Wildlife Theater. During these sessions, Rosie often grooms Jahangard, meticulously scanning his legs and arms with her black fingers, as the two men shoot the bull. "It's really comforting except when she pulls hairs out," Jahangard says.

What makes Rosie a males-only club is that only men can be assigned to her as trainers or backups. The thinking behind this has to do with how a female baboon would live in the wild. There, she would readily submit to male baboons, who are twice as big as females. The boys also have much bigger canines, some as big as a leopard's, and they will sink those into any uppity girl baboon. In the wild, male baboons are vagabonds. They leave their birth group when they reach sexual maturity and join up with another troupe, only to strike out again eventually. The females are stay-at-home types, remaining in their birth troupe their entire life. Girl baboons inherit their rank from their mother, so they don't scrap as often or as intensely as males do. However, they will readily battle any female from outside their troupe.

Instinctively, Rosie is inclined to challenge new female trainers, whom she's likely to view as coming from outside her troupe. She does this either by ignoring them or—worse—attacking them. Though she's had her canines pulled, Rosie's still got a mouthful of teeth, not to mention freakish strength. When Gary Wilson took over the school in 1985, Rosie still had female trainers. She got feisty toward one and eventually sank her teeth into the student's hands. Wilson says the student looked as if she had slit her wrists. After that, Wilson decided it would be only male trainers for Rosie. With men, Rosie is all lady, as in the throwback, kid-gloved kind. She doesn't mind that her male trainers come and go. She accepts each one of them as the boss, even a twenty-year-old like Jahangard.

Olive, the teaching zoo's other olive baboon, is a slightly different story. While Olive also requires a male trainer, she can have female backups. The thinking is that Olive is not as confident as Rosie and therefore would not be as likely to threaten a female backup. Rosie was born in captivity and handled by humans from an early age. Olive was captured from the wild and had a youth like a Charles Dickens character. She was caught illegally along with her mother in Africa. Olive was confiscated. The good intentions went awry as the baby baboon was separated from her mother and then housed alone during her impressionable youth. Consequently, Olive, now sixteen, can be very neurotic, shaking her den box, banging her head against it, self-mutilating, and pulling out fur. Her cage is littered with toys, mirrors, pieces of fabric—anything to keep her occupied. Still, the school doesn't risk giving her a female trainer. On the rare occasion that her female backups have asked her anything, Olive has ignored them. She really has eyes only for Jenkins. "Primates in general are incredibly chauvinistic," says Kevin Keith, an EATM grad and one of the few professional baboon trainers in the country.

Keith says it's not out of the question for a woman to train baboons. He knows of a female trainer who even works with male baboons. She is experienced, has an established relationship with the baboons, and is tough, he says. A female trainer can even have some advantages. "She can flirt with them," he says. At EATM, Goblin, the hamadryas baboon, has all female trainers, but that's because students work with her through the cage. She does not come out for walks like Rosie and Olive.

Pairing male trainers with Rosie and Olive is one of the many examples of how EATM students are taught to consider an animal's instincts and behavior. Students always hold macaws so they are above the birds, because in the parrot world whoever is higher is dominant. When students walk Julietta, the emu, three or so students trot alongside her. An emu is a herd animal and will stay with a group. Should Rosie or Olive ever challenge Jenkins or Jahangard, they are taught to grab the monkeys, push them to the ground, and hold them there—essentially what a dominant male baboon would do. Neither has had to do that so far. Keith takes it one step further at his facility. On the rare occasion that one of his baboons seriously challenges him, Keith will hold his baboon down and bite it on the back of the neck just enough to remind it he's in charge, as baboons do to each other. "I had to learn to hold my cheeks back when I bit. I bit my own cheeks." He rarely does it, but when he does, the bite works, he says. "You have to have that dominance or you are done."

Rosie and Olive aren't the only animals at the teaching zoo to whom gender matters. Neither C.J., the coyote, nor Legend, the wolf, cares for men. C.J. dislikes them intensely. She so dislikes Hyde, the towering teenager, that whenever he is anywhere near her cage, she barks and barks and barks. Julietta is with C.J., and decidedly doesn't like men. Once, as a group of students were getting her out for a walk, the emu swung her fuzzy blue head at Jenkins as if it were a club. As everyone collectively gasped, Jenkins coolly stepped close and grabbed the bird's long neck, which immediately calmed her down, and then said matter-of-factly, "She hates me." On the flip side, Ozz, the Geoffrey tamarin, prefers men.

Because there are always so few male students at EATM, chances are good that if you are a man you will be assigned to a baboon. You won't even have to score top grades or perfect attendance to get these high-status animals, unlike students who vie for Schmoo. However, Jenkins and Jahangard happen to be A students. Jenkins has clean-cut good looks (a style he undermines occasionally by shaving his head), bright eyes, and a high-pitched honk of a laugh. Jahangard has a long jutting jaw, dark curly hair, and rounded ears that stick out a bit. He's got the lean, muscular physique of a wrestler. He's easygoing, but his female classmates say he's got some ego.

Jahangard knew he wanted a baboon from the get-go, but there were originally five guys in his class. He had a little competition at least. Then two guys dropped out before school and a third not long after classes started. Jahangard was assured a baboon. Luckily, Jahangard and Jenkins did not have their hearts set on the same one. Jenkins, who has a psychology degree from UC Davis, was drawn to Olive's troubled personality. Jahangard just wanted to train behaviors, not psychoanalyze neuroses. Rosie was the one for him. She knows between seventy and eighty commands, and her standing flip will make you dizzy.

The fact that college students learn to work hands-on with the baboons here is a unique opportunity. There are few places to learn on the job, because only a small number of zoos or facilities bring their baboons out. Keith got his first baboon experience at EATM working with Balentino, a male who used to live at the teaching zoo. Keith was petrified of him. "It's like a race car driver. You might be fearful to get in the car but you're so fascinated by it." When Keith went in his cage, he'd square his shoulders and make his six-foot-five-inch frame look even bigger. Balentino would scream at him. Keith thought it was aggression. In fact, Keith was scaring the bejesus out of the baboon. He had to relax his posture.

After EATM, he worked at the animal show at Universal Studios and then at Six Flags Marine World in Vallejo, where he trained chimps. He gravitated back to the underappreciated baboons, eventually creating his own facility where he keeps four baboons and a mandrill, which he walks on a leash. Keith is one of a handful of baboon trainers nationwide. No one wants to work with the baboons much, he says. "They have red butts. They are aggressive." He not only likes the big monkeys, but sees his baboon facility as a wise business move. As pressure grows from animal rights organizations to not use chimps in entertainment, Keith expects demand for trained baboons to grow. They are quick thinkers, Keith says, and can work their whole life, unlike chimpanzees, which have to be retired when they become sexually mature at around ten. "[Baboons] are more socially intelligent," he says. "They always want to be in a group. They are like dogs. As long as you treat them in a positive manner, they love to be trained."

All the baboon walking makes for a busy day, especially for Jenkins. He gets Olive out at least once, usually twice, for walks. He is a backup for each of Rosie's two walks with Jahangard. Also, he has to follow along when Jahangard gets Rosie out for education shows in Wildlife Theatre. As the backup, Jenkins mostly runs interference, keeping onlookers from getting too close. He carries a cane, which is used primarily as a visual barricade to overeager humans, typically children, who often mistake Olive or Rosie for dogs. The backup is also there to keep passersby away from the trainer. The rule is nobody touches either trainer when they have Rosie or Olive out; you shouldn't even raise your voice to them. Either baboon might see that as a threat to their main man.

Olive's and Rosie's walks are not so much constitutionals as a change of pace in their captive lives. The promenades also keep them comfortable with being outside their cages. When Olive strolls with Jenkins, she is very alert, looking everywhere, examining the ground for food, keeping a constant eye on Jenkins. She pauses to smell bushes or to tuck a piece of bamboo in her cheek pocket. Rosie, on the other hand, is businesslike, almost blasé.

Both men keep one eye on their baboon and one eye on the world around them, scanning the horizon for anything that might upset their monkeys. For example, Rosie hates things that flap in the wind. Should something unsettle her, Jahangard will give her commands, such as asking her to take his hand, in order to calm her and get her focused on him. He does the same if Rosie shows any flicker of aggression—say if she eye-flashes a female student. This morning she is lovey-dovey. Jahangard asks for a kiss, and Rosie complies with a soft peck on his cheek. She drops from the bench and hunches on the ground by Jahangard's feet where she begins methodically sorting through the dark hair on his legs.

Rosie knows so many commands that it is hard to come up with new ones to teach her. For the winter semester, Jahangard will train her to voluntarily take an injection. He had thought of pricking her in her pink pillowy bottom, but the sciatic nerve is there. That means he'll have to prick her in the thigh. To do that, he'll have to face her, which makes him a little nervous. "All she has to do is bite me in the face." Rosie has rejoined Jahangard on the bench. He gooses her thigh to show how he's starting the injection training. Rosie starts a little and pops her bright eyes. This summer he trained her to let him reach in her cheek pouch. He demonstrates that too, poking his index finger in her mouth and prodding about while the monkey stares off in the distance.

Jahangard also trained her to climb atop the school van and leap into his arms like a damsel in distress. He gets up to show me. Rosie bounds on the hood. Jahangard thrusts one of his legs out. He opens his arms to the waiting baboon. She jumps easily into his waiting embrace, her black hands outstretched, her tail flying up behind her. Her feet land on his thigh. Her arms snug the young man around his neck. For a moment, human and baboon clasp like lovers.

Nearby, the jingle of female voices rises as EATM students file into Zoo 1. The day charges ahead. The men turn to walk Rosie back to her cage, the baboon following along on all fours. As the threesome ambles down the front road, Jenkins calls in his clear voice, "Baboon here!"

December

Despite the constant rush of life—the feeding, the cleaning, the gossiping, the studying—death is always in the wings at EATM. The students must learn to say good-bye for good. In early December, Mama Dolly, the doddering, ancient matriarch of the sheep, quits eating. Dr. Peddie holds a stethoscope to her stomach and listens for a sound like fingers rubbing against a wall. It's not there. That means the fourth chamber of her stomach has quit working. She is slowly starving to death. The vet rules that her time has come.

She's already lasted longer than anyone had expected. When the school bought six black-faced Suffolk lambs, the breeder threw in Mama. She could no longer breed but the ancient sheep could still be a leader. All grown up, the small flock of sheep still bleat hopelessly whenever Mama leaves the pen for a show or a training session.

The dowager arrived with an aging digestive tract. The stomachs of hoofstock churn and churn as they brew hay into a nutritional soup. Their guts eventually wear out from all that churning and brewing. As Mama's tummy failed, she has filled with gas, so much so the sheep has become noticeably lopsided. If you push her bulging side, Mama burps.

One early morning, Dr. Peddie again becomes a reluctant angel of death. CVP, Mama's student trainer and the crack pigeon puller, begs off death for once. Kristin Gieseker-Hopkins, an exuberant blond Texan who goes by the name of Kage, volunteers to take her place. She is the trainer of Sadie, the biggest of the small flock and the one that picks on the other sheep. Mama usually won't let anyone other than CVP halter her, but Kage thinks she has a shot; Mama knows her.

Turns out that Mama can hardly move. Kage and another second year have to lift Mama and half carry her out of the pen. Dr. Peddie warns them: when hoofstock are put down, there's nothing peaceful about it. They are given the initial sedative standing up. The animals sometimes go rigid, as if they've been shocked, and then collapse dramatically. Last year, when Dr. Peddie put down Bob, the hugely popular water buffalo, his great bulk not only hit the deck like a boulder, but he bellowed and cried.

Standing near Schmoo's enclosure, Dr. Peddie injects Mama with an overdose of a sedative. The queen sheep does not make much noise though she goes down hard. Her head lands in Kage's lap. She gurgles. That's unnerving, but what really starts to bother Kage is that the four sheep, including her Sadie, quietly watch the proceedings like witnesses at an execution. They never turn their coal black faces away. Their eyes, with those horizontal slashes of pupil, stare. Kage looks down and notices that Mama has dribbled a yellowish-green stain on her gym shoe.

It's Christmastime. This week, the students are playing Secret Santa, leaving little surprise gift bags on one another's desks. They have their pictures taken with Clarence, the Galapagos tortoise, whose shell is festooned with big red bows. Mostly, though, they fret. The holidays are lost to the nail biting of finals. In fact, the school's official holiday party is scheduled for mid-January, when the students can actually enjoy it.

For the Diversity final first years have to memorize over two hundred animals from sponges to storks. One morning, I find the metal picnic tables near the front gate full of first years flipping through handfuls of cards, their eyes scanning back and forth. There's much nervous tittering until a tall redhead snaps, "Everyone shut up!" One student, teetering on the edge of flunking Diversity, has broken out in hives. Linda Castaneda has bloodshot eyes. Sara Stresky, who rescued animals during the fire, has a migraine. Diversity is not her problem; she already dropped that class because she was failing. Her problem is Dr. Peddie's Anatomy and Physiology class. She needs a high score on the final to pass the class. For the test, she tells me, she'll have to take her meds for an attention deficit disorder.

The second years, as usual, don't seem nearly as anxious, though they have tests too. Most of the animals oblige and Training finals go swimmingly. Goblin, the baboon, presses the side of her head against her cage while Brunelli slides a thermometer into the baboon's ear. Brunelli's aim is solid this time. The only hitch is the thermometer doesn't seem to work. Amber Cavett was overambitious when she dreamed of training Samburu, the cranky caracal, to walk on a leash, but she shows Wilson she has taught the cat to present his front paws for a nail clipping. April Matott instructs Hamilton, the Yucatán miniature pig, to stay put while she measures his not-so-miniature hooked tusks. They all get A's.

Jena Anderson, Nick's trainer, only hooked the mini-horse up to the cart last week, but he gamely pulls it for her final. Though millions of horses have been taught to tow countless carts, it's still an amazing accomplishment. Anderson has convincd a prey animal to pull something it can't see. Anderson scores an A. If she was graded on his weight reduction, the score might be lower. Despite her best efforts, Nick is still a fatty. After Nick and Whiz recently moved to a new corral, Anderson tried again to tie the llama's food basket out of the horse's reach. Before long, Nick managed to knock Whiz's food out of the basket with his muzzle. Anderson gives up. The horse wins the battle of the bulge.

Likewise, after a semester-long effort, Mohelnitzky has not gotten to the bottom of Abbey's stomach troubles. Neither acidophilus nor Zantac makes a lick of difference; the pooch still pukes bile most mornings. Mohelnitzky did train the dog to zip over an A-frame, but Abbey doesn't do as well weaving through a series of poles stuck in the ground, so Mohelnitzky gets a B for her final.

A few animals do not cooperate, reminding these students of their fundamental wildness. On this sunny morning, Kage rouses Todd, the albino red fox, from a nap for her final. Big mistake: Todd, her beloved canine, does not like to be woken from a snooze. At first, he won't get off the shelf at the back of his cage. "That's unusual," Kage says. Still, her final gets off to a good start. He gets on a scale as told, and weighs in at 18.2 pounds. Todd wags his fluffy tail at Wilson, who stands outside the cage, and lets Kage stick her finger covered with toothpaste in his mouth. It's when she gets out nail clippers that everything slides south in a hurry.

When she tries to clip his toenails, Todd bares his teeth at her, even snaps close to her cheeks. "Cut it out," Kage barks and stands her ground. "What is your issue?" A professional trainer might step back, but Kage's grade is on the line. Wilson is watching. She runs the fox through a series of commands. Todd settles down. The fox lets her take a back foot while he stares kind of haplessly at Wilson. "It's so hard to be a fox named Todd," Kage says. Her gamble pays off. Wilson gives her a 100.

A little later, VanHollebeke steps into Starsky's cage and plops down on the berm of concrete. The cavy darts to a far corner of his cage, his little white skirt quivering. Starsky ducks behind a plastic igloo-shaped doghouse covered with pine boughs. VanHollebeke's oversized jacket rustles as she stretches out her arm. She holds her long stick with a spoon taped onto the end. In the spoon is a small pile of alfalfa pellets.

"Come here, handsome," VanHollebeke calls. "It's okay."

She lets out a big sigh. Since the fire, she's regained Starsky's trust, inching her way back into his cage, but you can't tell that just now. The cavy takes a baby step out from behind the igloo, one wide eye trained on us. His little body is rigid like a sculpture. We wait. He waits. VanHollebeke holds the spoon. The humans and the cavy all stand stock still. Around us life pulses. Buttercup, the badger, burrows through a mound of shredded paper with her flipperlike paws. Rosie, the baboon, lounges on her back atop her den box. Hudson, the beaver, strips bark off a bough with his orange teeth. Starsky looks as if a taxidermist got ahold of him, except for the faintest twittering of his whiskers.

"It's so very scary," VanHollebeke says. "I know, bubba." She shakes her head and sighs again. "He did it this morning." The standoff continues. To the cavy's thinking, it's a matter of life and death; for VanHollebeke, only a grade is at stake.

"I'm ready to quit when you are," she snaps. VanHollebeke stands up and steps forward to rake up the pellets she accidentally spilled from the spoon. In one quick hop, Starsky is back behind the igloo. Van Hollebeke seems on the verge of losing her famous temper, and if she wasn't in the cage with the scaredy-cat cavy, she might.

"Ninety-five," Wilson says. "I'll give you the benefit of the doubt."

"I'm determined to touch him before I leave this place," VanHollebeke says and purses her lips. Starsky pokes his head out from behind the igloo again, sees VanHollebeke's figure looming, and vanishes.

* * *

That afternoon in Zoo 2, the first years sit in the dark with pens in hand. They take a collective deep breath as image after image flashes by. They chew their lips, tap their pens against their forehead, squirm in their seats. An hour and a half later, they stumble out into the sunshine, squinting. Now, they all breathe a sigh of relief. "I'm going to have a slide-burning party," a first year announces.

The second-guessing begins. "What was the blue bird with the funny head?" one first year asks, holding her hands atop her head and waggling her fingers to demonstrate its crown of feathers. "What kind of rhino was that," Castaneda asks, "a black rhino or an Indian rhino?" "Black," say a few. "Indian," counter others. "It could have been worse," Castaneda says, her eyes pinched from too little sleep. "He could have given us all four rhinos. He only gave us one."

Castaneda guesses that she missed eight questions, if she counts the rhino. That gives her a B, but good enough to keep her A in the class, which preserves her perfect record. If only her classmates could be as diligent! She turns to show me the first years' most recent gaffe. My eyes scan the back of her sweatshirt. Each class creates a design for its sweatshirt. The first years went a little philosophical, using a quote from Gandhi, but they misspelled his name; the *h* is missing. Castaneda, the driven daughter of immigrants, frowns over her shoulder at me.

I follow Susan Patch back inside Zoo 2. She's going to give Missy, her rat, the last practice run on her Wild West–themed maze. The test is tomorrow. Missy has a black head with dark pink ears and a stripe down the back of her shiny coat. Patch plops Missy on her shoulder. She hasn't just trained her rat to run a maze. She's created a piece of theater. The maze is a mini multilevel stage set. The top level has a painted backdrop of snowcapped mountains; the next level resembles the Grand Canyon; the level below that has a diorama of cactuses. On the bottom is a little saloon, complete with swinging doors, Patch created with popsicle sticks.

She made up a story for the maze—a kind of classic Western, but starring a girl rat. As Patch tells it, Missy has to brave a mine shaft (a toilet paper tube), cross a precarious rope bridge over a gorge, climb down a ladder to the desert below, where she must wind her way through cactuses (weave poles) and forge the River Plastico (a plastic container of water). When Missy gets to town, night has fallen. "What would any rat do?" Patch asks. "Go to the local bar." Missy pushes open swinging doors, tosses back a shot at the bar, and plays poker. Patch trained Missy to press her nose to her thumb, which Patch always sets atop the Queen of Hearts. Because she's a wily rodent, Patch explains, Missy cheats at cards, and thus "makes a hasty exit." The rat opens a little window and dives through it.

Patch has trained Missy on the maze twice a day, every day for three weeks, rewarding her with dried banana chips and sunflower seeds split in half. Like a coach, Patch warms Missy up with a massage and then plenty of chatter. As the rat noses around Patch's shoulders, ducking under her ponytail, and climbing in and out of her hood, the first year prattles on. "Are you ready? Are you ready?" Then she sets her hand on the top of the maze and trills, "Go, Missy!" Missy gamely trots the length of her arm and begins her journey through the Wild West. She takes her time going over the gorge. "Come on, you slug," Patch says. Missy stops amid the cactus weave poles to bathe herself, rubbing her little lavender-pink paws over her face. "No baths," Patch says. "Gary says they bathe when they are nervous." When Missy gets to the Rio Plastico, she takes a sip before ambling over the little bridge. She jumps down to the saloon and stalls by the swinging doors, then ignores the handful of playing cards Patch holds up to her. Patch grabs Missy and deposits the rat back on her shoulder.

"Are you ready? Are you ready? We'll be faster tomorrow. You betcha."

By the third run Missy begins to pick up a head of steam, and on the fourth go she smokes the maze, zipping over the rope bridge and dashing through the weave poles, her little head bobbing left, right, left, right. She bounds over the Rio Plastico, throws open the saloon's swinging doors with her little pink paws, dips her nose into the thimble Patch is using for a shot glass, picks the high card, and bolts out the window.

Patch, wanting to end the practice on a high note, calls it a day. Back in her cage, Missy thirstily takes a pull on her water bottle, her pink nose twitching all the while. "If anything happens to her overnight," Patch says, looking at Missy, "I'm screwed."

The next morning, I drive to EATM along an empty highway under a full moon. The road curves gently like a wide river through the burned, bald hillsides. They are black no more. The scorched landscape, at least at this early hour, glitters like a forgotten family heirloom. Time and winter's cool moistness have burnished the landscape bronze. Under the thin dawn light, the charred, bent-over cactuses resemble gilded filigree. Swaths of fire retardant are the milky green of oxidizing copper.

EATM students zoom into the parking lot, grab their backpacks, and rush for the gate. The vegan from Colorado runs past me as I open the tall chain-link gate. She mutters hello as she goes by. She's trailing the main pack of students by about thirty seconds—thirty seconds that could make her tardy; thirty seconds that could keep her from training the animal she's set her heart on. The minutes, the seconds, pass with such consequence at this early hour.

Every Wednesday morning all the students and most of the staff meet. It's the one chance each week to pass on zoo news to pretty much everyone at once. So, as soon as the hoses are coiled and rakes set aside, all the students file into Zoo 2. The second years sit close together, massage each other, and pass boxes of slick doughnuts among themselves. The first years seem by comparison sunken-eyed and harried. A number of them have their rats in tow, enclosed in little cages they clutch like pocketbooks. The maze test is this afternoon. Michlyn Hines announces that the zoo's truck is out of oil. The mother of one of the second years is dying of cancer, and Hines needs student volunteers to cover the absent student's cleaning shifts. A young water buffalo is moving in next week. It's coming with a zebu. Taj, the Bengal tiger, has runny poop. Wendell, the pygmy goat, has blood in his urine.

As usual, Brunelli has one, two, or three announcements between student council, the yearbook, and her fund-raising schemes. Kage, nicknamed The Dish Nazi, stands up and complains that people are *still* leaving their dirty bowls in Nutrition. Chris Jenkins rises to say a broken bulb was found in the enclosure with Sally, the snapping turtle; the zoo is almost out of crickets; and don't touch the baby sand boas. They'll be for sale soon.

The day unspools like any other day at the teaching zoo, except that it isn't for about a dozen first years. Today is a pivotal day for them—their first chance to demonstrate their training chops for a grade. The test is also their first taste of what it's like to have their fate hang on an animal, in this case a rat. Come mid-afternoon, Zoo 2 buzzes with cheerleading. With just a few minutes left before the rat maze test, first years coach their rodents through practice runs. "Jump it!" "Cross it!" "Rope!" Kristy Marson, one of the most enthusiastic students in her class, is the loudest. Her black ponytail bobs furiously as her rat bounds into a makeshift mini-tram and rides it to a lower level in the maze. Only Marissa Williams is quiet. Williams has big pretty eyes and a dancer's figure and grace. She watches dejectedly as her rat dangles from a rope ladder. It won't climb. Her first two rats became sick, sneezing so hard they got bloody noses. She got a third rat, but it became ill, too. Dr. Peddie prescribed antibiotics, and the rats recovered but not soon enough. Williams only started training one last night.

At the top of her maze is a little TV set. The rat is supposed to put his head in it, so that his little furry face fills the picture tube. He pokes his pointed pink nose in, twitters his whiskers, and then starts to rub his paws over his face. He's taking a bath. Williams sighs. Wilson teaches that if an animal gets a behavior wrong, it's the trainer's fault, not the animal's. Williams's expression says she's having a hard time not blaming her rat, or at least her luck.

The rules for the rat maze test are simple, as Wilson explains: Each student has three minutes. If the rat runs the maze more than once, Wilson will grade the best run. He will deduct for touching the rat or feeding it while in the maze. It doesn't matter if you use verbal cues or not. They will go in alphabetical order, but you can volunteer to go earlier, which is what Patch does. She will be tenth.

The first rat runs a bare-bones maze of foam board with no problem and pops merrily out a window at the end. The second rat, a plump one, stalls in the middle of his run. The third student pulls a rat from her hoodie, and it blasts through an elaborate model prison, climbing a mini razor-wire fence and jumping into the back of a getaway truck with the word FREE written on the side. Everyone claps. The next rat—a rosy, hairless number—runs around like a windup toy, here, there, everywhere but into his maze. As her grade plummets, the student can't help laughing. Wilson cracks up too.

Then Marson gets up. "Sorry, I have to be loud for my rat," she says. As Marson bounces and chants, her rat races through the maze, even swimming across a little pool, her pink paws doing a kind of breast stroke. After her rat blasts through a sixth time with the three-minute clock yet to run out, Wilson says, "You can stop now. Your grade can't get any better."

"My rat just peed all over me," another student announces, and leaves the classroom. A few more students get up, including one of the few guys in the class. With a knit cap pulled down over his eyebrows, he stands still and silently watches as his rat bobs through a clear plastic maze. Finally, Patch is up. She steps to the front of the room.

"Sorry, I talk to my rat," she says.

"Susan, you have whole conversations with your rat," Wilson responds.

As Patch places her maze on the table, the mountain backdrop at the top comes loose. Marson volunteers to hold it in place. Missy scampers back and forth along Patch's shoulders and under her ponytail. "It's going to be good. I'm going to get you. I think you can do it, baby girl. Okeydokey, okeydokey," Patch chatters, then sets her hand on the right side of the maze. Missy scampers down, but freezes when she sees Marson, her face looming over the toilet paper roll mine shaft. Marson ducks out of sight, but Missy is spooked.

Patch starts over. This time Missy ignores Marson and trots through the mine shaft but halfheartedly. The rat takes her time on the rope bridge, meanders through the cactus weave poles, pads around the bridge, freezes outside the swinging doors to the saloon, and rubs her little paws worriedly over her face. Missy's clearly not feeling very cowgirl. "I'm going to beg," Patch says. "Don't do this to me. Don't take a bath. We don't have time for this. I'm going to poke you." Patch picks her up for a second run, but Wilson calls time. Patch plops Missy back on her shoulder. "That was not real fast, not real slick, not fast or slick, but it was all right," Patch says.

After Patch, a student gets up with a fancy maze in the shape of a two-masted schooner. The rat's performance is less impressive; it won't even ring the bell. The next student pleads with her rodent, "Don't you dare do this to me." Another says, "Come on, momma wants to pass this class." The last student to go is Williams. Her rat manages a bit of the maze, even scoots up some of the rope, but then, as usual, bathes by the little TV set. Williams frowns and sighs. "I'm not one to beg rats," she says. She waits out the clock basically. When Wilson calls time she dejectedly picks up her rat and walks away, leaving her maze behind.

"Kristy, you won," someone calls.

"I didn't win," she says. "There's nothing to win."

"You ruined the curve," Wilson says.

It's always something with Schmoo. Her pool needs fixing or she's not eating. Over the past five days, she's gotten toothy, even bitey. On Saturday, Schmoo totally ignored one of her student trainers during a show. A staff member had to step in. Two days ago she clamped on to one of her student trainers' thighs. "I was stunned stupid that I was that stupid," she tells me. "You're always expecting the unexpected to happen, but when it does you're shocked." The bruise on her leg looks like she ran into a coffee table. This morning Schmoo, her black teeth bared, charged another student trainer twice and chased her a good twenty feet. This student now keeps six feet between her and the sea lion.

In addition to the aggression, Schmoo has also stopped eating in the morning. Yesterday she turned her nose up at all fish until 3:30 P.M. Now Wilson slumps in a collapsible chair in the shade near her cage. His brim is pulled low. He was up until 2 A.M. writing the Diversity final. The Schmoo girls, the sea lion's four second-year trainers, stand nearby. All four have various shades of blond hair. The object of their consternation sits in her corner, her long whiskers glowing in the late afternoon sun.

Five human brains join forces to decipher one sea lion brain. This is the crux of training, trying to think like an animal, which turns out to be not so easy for humans. Maybe it's her pool. It's dirty enough to make even the most levelheaded sea lion peevish. The swirling water is a murky brown, and when Schmoo hops out of it, the smell of sewage follows her. The filter, which was meant for a swimming pool, just isn't up to all the poop and sand, but the school can't afford another.

Wilson suggests using rewards other than food, such as pieces of ice or spray from a mister bottle—in behaviorspeak "a secondary reinforcer." He explains that most places work with big, 500-pound male sea lions that are very confident and very food-driven, so they are not as complicated as the comparatively petite Schmoo. Wilson and the second years chew on this for a while, but no conclusions are reached. He leaves, and the trainers get the recalcitrant sea lion out of her roomy enclosure. She happily waddles out and starts working with one of the second years, the one she bared her teeth at this morning. Schmoo salutes, sticks her tongue out, and flips her tail up. Then the trainer turns to me and asks, "Ever been kissed by a sea lion?"

"No," I answer, and before I know it, the trainer points Schmoo in my direction and commands "Kiss!" She's scooting my way, her near-blind eyes fixed on me. The conversation I've just heard about her charging and biting races through my head. I close my eyes extra tight and purse my lips so they are as far away from my face as possible. Our kiss is more Eskimo than anything. We bump noses clumsily. I feel her coarse whiskers against my cheeks. She exhales and her sea lion breath rushes up my nose. I hear a light grunt, and she's gone. My face is intact, though I can't help checking my nose with one hand. It's moist on the end. I smile long and hard and, for some reason, blink back tears.

The training session continues with each of the four second years taking turns. One second year kicks a soccer ball at Schmoo, and the sea lion bounces it back with her nose. The student gets a cart out. Schmoo hops aboard and waves one flipper as the second year pulls it around the middle of Hoofstock. Both camels watch the sea lion wheel by like a beauty queen in a parade. All seems normal. The last trainer, an especially cool-headed second year, orders the sea lion into the large box to one side of her enclosure. Schmoo complies. The student steps into Schmoo's enclosure to reward the sea lion with fish. Then, as the student turns to leave, Schmoo waddles after her. The second year orders her back. Schmoo, all black teeth, charges. The rest of us gasp.

The student backs up out of the cage a good ten feet before Schmoo stops. "Relax!" she commands, and Schmoo does. As the second year walks the sea lion back to her box, Schmoo flashes those black teeth again right by the student's waist. The second year dodges Schmoo, keeps her composure, and orders Schmoo back into her box. The sea lion obliges as if nothing has happened. The student closes the gate. The four Schmoo girls shake their heads, then walk to Nutrition to rinse the fish goo off their hands. I nervously touch the end of my nose, still damp from my sea lion kiss.

Before finals week is out, Brunelli breaks up with her actor boyfriend. She doesn't blame EATM but guesses the program hastened along the inevitable. She's relieved and even has someone in mind to date, a teller at her bank. Still, the day of the breakup, she asks another student to feed Rowdy, the skunk, and goes home upset. Sequoia, the deer, drags a second year, a young woman from Salt Lake City with a dry sense of humor, down the back road. The deer snaps the chain taut and the second year is briefly airborne. The student comes down hard, rips a wad of flesh off her elbow, and lets go. Sequoia's tail flashes as the deer bolts away. "Deer loose!" someone yells. Sequoia comes to an abrupt stop by the door to Hoofstock, where the school's gardener, the deer's favorite human, catches her.

Second years begin what they call turnovers, essentially the process of one student trainer passing their animal to the next student trainer. They demonstrate commands, detail likes and dislikes, show them how to make the animals' dinners. One afternoon, April Matott watches while another second year puts Savuti the hyena through his paces. To the command "Look cute!" he puts his front paws up on a den box and looks back over his shoulder. "You have to tell him he looks cute," the Savuti trainer tells Matott. You can train him to obey commands in one session, she says. The trick, she tells Matott, is your timing must always be precise.

There's a potluck lunch, during which the Secret Santas are revealed and more presents are exchanged. The strain of the week shows. Despite the thumping bass of a boom box, Chris Jenkins falls asleep at the back of the classroom with Abbey's leash in his hand. The dog naps at his feet. Castaneda's eyes are still bloodshot. Sara Stresky tells me she thinks she might have flunked Dr. Peddie's test. The morning cleaning ran over that day, and she arrived a half hour late for the test. She forgot to take her meds for her learning disorder. Will she be the next in her class to go?

Kage's week, which started with holding Mama Dolly while she was put down, never improved. After Todd, the fox, tried to bite her during her final, Benny, her capuchin monkey, nipped her. So did Alamo, the prairie dog. Finally, when she went to get Sadie out for a show, the sheep ran from Kage as if she were afraid for her life. Sadie won't even take food from Kage, which is unheard of with ever-hungry sheep. Wilson teaches that animals associate events that happen simultaneously or close together. If a dog is going through a doorway when a car backfires outside, the dog is likely to be afraid of doorways from then on. What went through Sadie's mind when she watched as Mama Dolly collapsed in Kage's arms? The sheep balks as Kage tries to lead her up the back road. Sadie digs in hoof after hoof.

Dr. Peddie

There are some things you should know about Dr. Jim Peddie. He is deaf in his left ear. He is a good storyteller, so much so that when he tells you the same story a second and even a third time, you don't mind. Nothing grosses him out. He can eat with gusto while discussing the components of tears: oil, water, and snot. He's all boy. He fishes. He loves cars, planes—basically, anything that moves.

Dr. Peddie has a beautiful, smart wife. He made a bundle. He was a vet to the stars. He's delivered baby elephants, removed a garden hose from a hyena's stomach, and brought black leopards back from the brink of death. He flew to Africa and helped paint a lion white. He stood in the wings of the Academy Awards, a blow dart pipe aimed at Bart the Bear, in case the 1,500-pound Kodiak onstage attacked Mike Myers.

He's lived an incredible life. Not a fairy-tale life—there have been some bumps along the way—but a big, rich, exciting life. He still can't believe it, and his voice can fill with wonder, as if he's no idea how it happened, his life. But he knows. True, this country boy from Pennsylvania is talented and personable, but plenty of people are. Dr. Peddie's put in long days for years and years. He's always kept farmers' hours—rising in the dark, often toiling well into the evening. He hasn't slept more than six hours in years. "I've worked ever since I can remember," he says.

Now, he says, he's tired—tired of paperwork, committee meetings, putting animals to sleep, and arguing. He looks fit, but decades of overwork have left him with a list of health problems longer than Schmoo's. He's got spastic colon, vertigo, migraine headaches, and carpal tunnel syndrome. Every day his thyroid destroys a little bit more of itself. It's time to slow down, he says. He wants to tinker with his collection of small antique engines, buy a boat, and travel with his wife. Still, I'm not entirely convinced. It's hard to believe that a man who idles at such a high rpm can slow his pace, even if he wants to.

We are sitting in a dimly lit Chinese restaurant in a low-slung shopping mall off the 101 in Oxnard. There's a soothing hush to the restaurant. Before us is an overwhelming spread: a stack of moo shu pancakes, a slick stir-fry, and a bowl of white rice. Dr. Peddie digs in. It's not long past 6 P.M. Dr. Peddie eats like a farmhand—voraciously and on the early side. Once when I ferried a box of barbecue takeout to his beach house in Ventura, I arrived at 6:45 P.M. to find him pacing his driveway like a ravenous big cat.

Dr. Peddie looks like a guy who would list splitting wood as one of his hobbies. He is tall, big in the shoulders, and a little broad of girth. He has a wide-open face that is easy to read. He purses his lips, scrunches his nose, even rolls his eyes. He sports oversized wire-rimmed glasses and favors baseball caps, but takes them off inside, as he has for dinner. He is one of the best dinner partners you can have. Not only does he have good stories to tell, though you might not want to hear the one about the constipated elephant while you're eating, but also he's interested in most everything, solicitous, says what's on his mind, and laughs heartily at your jokes. His only fault as a conversationalist is that he's prone to exaggerate. Someone he likes is the gold standard; someone he doesn't is the biggest asshole who ever lived. One of his oldest friends jokes that Dr. Peddie inflates all the numbers in his stories by at least 30 percent. His wife used to correct him as he embellished his tales but gave up.

This is his perhaps-somewhat-embellished story. Dr. Peddie set his sights on becoming a dairy vet early on. It was a logical choice for an ambitious boy in Williamsport, Pennsylvania. The town, a county seat, straddles the west branch of the Susquehanna River as it slips through the foothills of the Allegheny Mountains. Other than the surrounding terrain of deep woods, old mountains, and plentiful rivers, Williamsport didn't have a lot going for it then. The former lumber town was a town of small dreams, the vet says. Unemployment was high. Most jobs there were blue collar. Even though his family was middle class—his father was an airplane mechanic, his mother a teacher—Dr. Peddie grew up feeling that the wolf was at the door. That gnawing sense of impending destitution, combined with his mother's forever telling him to get busy, even if he was reading a book, made for a boy who became a driven, restless man.

"All of the years I've known him, he's either going ninety miles per hour or he's asleep," says his wife, Linda.

He was the first member of his family to go to a four-year college. He attended Cornell University for seven years: three years pre-vet, four years in vet school. His only sibling, a younger brother, followed the same path. Dr. Peddie's central accomplishment at Cornell, which he still boasts about regularly, was snagging the only female student in the vet class of '65, the then Linda Reeve. Linda hailed from a farm town on the far end of Long Island. At first, she didn't know what to make of the unsophisticated young man from Pennsylvania. He could be so candid it hurt, Linda says. She did notice his hands. During dissections, he moved them with such sureness and ease. On their first date, they stopped for gas, and he swapped fishing stories with the attendant. She liked the way that he could "scratch and spit with the best of them."

Dr. Peddie saw himself as the future James Herriot of central Pennsylvania, but Linda changed this. As he puts it, she told him "there ain't no way in hell that I'm going to work on cows." Each had spent a glorious summer in California. In 1965, they married during spring break, graduated, packed their car, and turned it west. They had hardly settled in when Dr. Peddie was drafted. He spent his two-year tour of duty not in the jungles of Vietnam but in the walk-in freezers of Fort Lee, Virginia. He oversaw the base's food safety. Though he stayed stateside, Dr. Peddie still got hurt. He pulled up along a tank on a firing range just as it kaboomed. His left ear instantly began ringing and has never stopped.

Discharged, he and Linda returned again to California with the first of their two daughters in tow. They both joined the staff at Conejo Valley Veterinary Clinic in Thousand Oaks. Bob Miller, now a famous equine specialist, had started the practice in his garage, using an ironing board for an operating table, Linda says. By 1968, when the Peddies came on board, the practice called a building on Thousand Oaks's main street home. Though Thousands Oaks has since morphed into a hopping commuter burg with SUVs barreling from shopping plaza to shopping plaza, in the late sixties it was still an outpost, one where you might see a lion chained along the main drag or maybe even an elephant out for a stroll.

Just down the street from the clinic was Jungleland. The wild animal park was started in 1927 by the animal importer Louis Goebels. In its heyday, Jungleland drew crowds from LA to see its performing lions, tigers, elephants, and chimps. Circuses wintered on the park's grounds. Lions' roars and peacocks' screeches roused locals from their dreams. Schoolchildren were kept home for the day when six mountain lions escaped from the park. By the 1960s, the park's glory days were behind it, but it was still open to the public and the animals continued to work in movies and television. The trainers Hubert Wells and Wally Ross worked there as well as old-timers like Mabel Stark, the heavily scarred tiger trainer who had been a Ringling Bros. star long before.

Conejo Valley Vet cared for the animals, though none of the vets was trained for exotics. Vet schools then rarely broached the subject of exotic animal medicine. Dr. Murray Fowler at the University of California, Davis, offered the first course ever in exotic animal medicine in 1967. Until then, how to sew up a torn elephant trunk or treat a constipated cougar was acquired on the job and passed down by word of mouth.

Before Jungleland, the most unusual animal the Peddies had worked on was a wild boar in vet school. "We had zippo for confidence," Linda says. "Miller's approach was, well, 'What kind of animal is it like?' He thought if we don't try to help them, who will?" Dr. Peddie answered an emergency call from Jungleland about a sick black leopard. He found not a sick cat but a nearly dead cat. Her uterus had ruptured, dumping her three unborn cubs into her abdominal cavity. He operated on her at the clinic, saving her and the cubs. "I was wetting my pants," he says. "Nobody ever said anything about leopards in vet school, but, hell, I did this."

Dr. Peddie loved the adrenaline rush of working with exotics, the thinking on his feet, the pioneering aspect of it. Working on exotics was a tricky business then. There were few sedatives to choose from. Owners rarely knew the weight of their animals. If you gave an animal too much anesthetic, you could kill it; too little, the animal could kill you. Dr. Peddie once had a tiger sit up on the operating table. "I was always pooping in my pants when I worked on these animals."

His confidence grew quickly. He anesthetized Jungleland's mandrill, which he was treating for lymphatic leukemia. After the treatment was completed, he and Miller loaded the doped mandrill into a pickup truck. Dr. Peddie sat the mandrill on his lap, propping the animal up in the passenger window so passing cars would see his exotically colored face. They drove to his house, leaned the mandrill up against the door, rang the doorbell, and hid out of sight. When Linda answered, she took one look and said, "Whose idea of a joke is this?" It wasn't the effect the vets were after. They took the mandrill over to his neighbors'. That woman opened her door and screamed. Satisfied, the vets returned to driving the mandrill around town.

By the 1960s the animals at Jungleland did not always get the best food, if they even got enough. Consequently, the Conejo vets saw cases they had only read about in textbooks. The Asiatic deer got oleander poisoning from chewing cuttings near their cage. All the tigers fell into Rip Van Winkle–like slumbers. The cats had dined on a horse, including its liver, that had been euthanized with barbiturates. Four of the tigers died. The survivors snoozed for as long as four days.

In 1969, Jungleland finally closed. One of its lions took a bite out of Jayne Mansfield's son while the movie star was posing for publicity shots. She sued for big bucks. The park's 1,800 animals were auctioned off. Still, there were plenty of exotic animals in the area. There were movie trainers and circuses scattered around. Hubert Wells had started his own company, Animal Actors of Hollywood. "I came to realize that a lot of these people were very decent people," Dr. Peddie says. "They were tremendously devoted to their animals. You always hear about these people beating the animals. It was 180 degrees from that."

Not long after he started EATM, Bill Brisby had asked Dr. Peddie to teach a class. Dr. Peddie thought the program sounded flaky and begged off. By 1977 he had changed his mind and began teaching there one night a week, lecturing on parrot nutrition or how to diagnose pinkeye, from 6 to 10 P.M. each Monday. His exhaustive essay tests quickly became notorious. His focus remained the practice, and he became a partner as well as business manager. He cranked out eight surgeries a day. Those hands that Linda Reeve fell for could work magic inside and out. Not only could he patch the animals back together, but they hardly had scars. Vet work is hard, physical work, especially surgery. What with the surgeries, working weekends, and never sleeping enough, Dr. Peddie wore out. His hands ached and tingled from carpal tunnel syndrome. He cashed in his share of the practice and signed on to teach full time and be the staff vet at EATM. He was forty-nine. The plan was to slow down and take vacations and holidays. He tried to but couldn't.

Gary Gero, a top Hollywood trainer who runs the show at Universal Studios, asked the vet to visit his facility once a month and give the animals general vet care. It was such good money, Dr. Peddie could not say no. In short order, word spread about Dr. Peddie among studio trainers, and the Peddies found themselves with a new business—a vet practice for movie and TV animals. They called it Drs. Peddie. Life sped up to double time, then triple time. They both worked seven days a week. Dr. Peddie taught his classes, cared for the teaching zoo, then drove off to tend Moose, the Jack Russell terrier on *Frasier*, or zipped over to Universal Studios to check on an orangutan.

This was as glamorous as vet work could get. Trainers solicited his opinion. He hung out on movie sets. He jetted to Australia to be on the set of *Babe 2*. He zigzagged around Southeast Asia with a trainer looking for an orangutan to use in the movie. "It was really stimulating," Jim said. "It's the people. They made my mind work. I'm a different person with them. My thought process kicks up a notch."

It was not so stimulating for his wife. Linda was stuck at home. She was mission control, piloting the business and the substantial paperwork. Every time an animal went on a foreign set, document upon document was needed. "If I went grocery shopping, when I got back there'd be twelve to fourteen messages on the machine," she says. "I gave up reading. I had no life outside of the practice. He was running around the countryside while I was at home doing the books."

In 1999 all of Linda's joints ballooned and throbbed. A year later Dr. Peddie's joints blew up too. He also developed vertigo that brought on spells of spinning that were so bad he threw up for hours. Though he'd had surgery, his carpal tunnel syndrome flared. His hands would fall asleep and he couldn't hold a syringe. Still, the vet couldn't stop, not even slow down. In 2001, between jobs, he zoomed off for his annual all-guys luxe fishing trip in Alaska. He came home to find that Drs. Peddie was no more.

"I shut down the business because I'd had it," Linda says. "I knew he was compromised as he was. It was ridiculous to go off on a fishing trip. He couldn't even tie a tie. I told him when he left, 'I'm shutting this down.' He loved to be a vet to the stars. It was a larger-than-life kind of thing. . . . I said, 'Jim, the animals you're dealing with know you're compromised. They know it inherently.' " Dr. Peddie was furious. "It was my identity," he says.

No matter how down to earth or confident Dr. Peddie is, he cares about status. He'll admit it. He was a big somebody and now, thanks to Linda, he wasn't. Luckily, he still had EATM. That same year, Dr. Peddie, somewhat to his surprise, ascended to the position of department chair of EATM. The program had a new dean, Brenda Shubert. She made big changes, which resulted in Gary Wilson's stepping down as EATM director. Shubert asked Dr. Peddie to take over. If he didn't, the college was going to bring in someone from outside of the program. Dr. Peddie did so reluctantly, he says, which caused a falling out between the vet and Wilson. They have been at loggerheads ever since. You will rarely even find them at the same social event. Few, if any, of the students know about this rift, because the tensions play out in staff meetings behind closed doors. The all-female staff, though, is keenly aware of the bad relations. You could write it off as a struggle between two alpha males, which is likely part of it, but both men have very different styles and very different personalities. Wilson is a thinker, a dreamer; the vet is a doer, Mr. Practical. Whatever the cause, the two men just do not agree, and Dr. Peddie is tired of arguing.

As head of the program, Dr. Peddie's been approachable, not a mythic character like Briz. He might be gruff with students on occasion, when he wearies of their various false alarms about the animals. For a man's man, he's at home with women, which is a good thing at EATM. He can be a bit fatherly, maybe because he has two daughters. He tells me students often ask him for medical advice, confiding in him that they have a rash or that it burns when they pee. "I draw the line at, 'Will you feel this lump?' "

Dr. Peddie, like most men, also enjoys the company of attractive young women. During my first visit to the school, I scanned the photos of second years posted in the student lounge and asked the vet, What was with all the babes? He acted surprised, looked closer at the photos, as if he'd never noticed. He'd noticed.

The vet still feels bad about Mama Dolly, he says, as we make a dent in the moo shu chicken and stir-fry. "It's a sense of failure," he says. A couple of other animals at the teaching zoo hover at death's door. George has some mysterious ailment that makes the little fox as dizzy as a drunk. Louie, the ancient prairie dog, who's paralyzed from the waist down, steadily declines. Schmoo always worries him. She seems fit, but given her epilepsy and age, she's fragile. "I hope I get out of here before she crashes," he says.

Dr. Peddie will retire this coming May. He will have taught twenty-six years. He tried to make his exit last year, but the college president talked him out of retiring. His wife was disappointed.

Just one more year, he promised her. Just one more year, he told the president. He's keeping his bargain, though the first years hope to change his mind. He's ordered a boat to be built up in Paso Robles. He and Linda plan to go to New Zealand. He's anointed his successor: Brenda Woodhouse. Next fall, she'll become EATM's first female director.

What will Dr. Peddie's leaving mean for EATM? The vet has taught at the school longer than anyone. Unlike the time when Briz retired, nowadays no one is worried about the fate of the school, though some staff members wonder how it will fare without the vet's business sense and his good relations with the dean. What will it mean for the vet? "It's scary, this idea of shutting it down," he says. Still, you can already sense a bit of detachment in Dr. Peddie and that his future lies elsewhere, outside of the teaching zoo's front gate. He hasn't even gotten to know many of the first years' names. Come next summer, there will be no more grading tests, no more college meetings, no more paperwork, no more putting animals down—just fishing, trips with his wife, tinkering with his boat, and more fishing. His only worry is, can he be a nobody?

Walking Big Cats

The few weeks between the end of the first semester and the start of the second is a long vacation for most Moorpark College students but not for the EATM students. While the holiday break shutters the campus, the teaching zoo hops. There aren't any classes, but students must still tend to their charges, and not even Christmas liberates them from scooping poop. The only student to get a vacation is the one EATM student who is an EATM student no more. As she feared, Sara Stresky did not pass Dr. Peddie's test. He takes her in his office to tell her that she is out. Dr. Peddie always has a box of tissues ready for these meetings. Stresky digs into the box, wipes away tears. She can no longer venture past the front office during the week. She can't pull on her EATM coat, sweatshirt, or T-shirt. She must wait for the weekend like the rest of the general public to stroll down the front road. There is one advantage to her new low status: on Saturday and Sunday, Stresky, like any visitor, can chatter away with the teaching zoo's residents to her heart's content.

Stresky is the semester's third casualty. A seond year, the reputed pathological liar, finally got the boot in December, after repeatedly reporting late. In the past EATM students mostly washed out or quit because of the physical demands. Now, the academic demands are more often the culprit. Dr. Peddie says students focus on their zoo duties at the expense of studying. That may be so, but perhaps the students are only doing as told. The staff constantly chants the mantra, "The animals come first."

During the holiday break, the long-awaited water buffalo, Walter, arrives with his bosom companion, not a zebu, but a Scottish highland cow named Dunny. The twosome move into a pen in Hoofstock along the front road. They stand side by side, roughhouse occasionally, and look exceptionally cute for bovines. Somehow, like two mischievous boys, they manage to turn a water faucet on and flood Hoofstock twice. Walter's young horns sprout from his rounded forehead. They are only half as long as his ears. Dunny's enormous bulk is covered with a dense, caramel-colored coat worthy of a woolly mammoth. Kaleb, the camel, leans his long neck over the corral fence and sucks on Dunny's coat. The Scottish cow just stands there.

For two weeks George the ailing Fennec Fox pads in dizzy circles, his head cocked as if listening for a faraway rustling with his two large ears. All his tests come back normal, neither a blood cell out of place nor a worrisome shadow on his wee brain. After his student trainers find him trembling uncontrollaby one early January morning, Dr. Peddie euthanizes the small fox and then sends his little body off for an autopsy, in case he had something contagious such as meningitis. Meanwhile, one of the teaching zoo's pigeons, a white one, is found dead. Somehow, amid the small flock, it starved. Some students cast an accusing eye at the pigeon's second-year keepers. This is the place to make mistakes, Dr. Peddie says. That is why it's called a teaching zoo.

Becki Brunelli recovered quickly from her breakup. She's had several dates, including the bank teller and a contestant on the television reality show *The Bachelor*. However, there is a new male in her life who causes her endless pain—Cain, the chattering lory. Brunelli is assigned to the small nectar feeder for next semester. To that end, she's begun handling Cain so they can get used to each other, but every time Brunelli takes the little, loquacious lory out of his cage, he pierces her with his needlelike beak over and over. Her hands and forearms are polka-dotted with bruises and punctures. It's a comeuppance for the heretofore unbitten Brunelli. This little red bird with green wings has made a confident, smart, beautiful, energetic woman second-guess herself. "For me, most of my animals have taken to me pretty well," she tells me. "Your ego wants to say, animals like me. Then you have an animal who doesn't take to you. And you realize it's not about you."

This three-week stretch is another chance for second years to head off on projects. Most stay close to the zoo and save their money (they often have to pay transportation, and room and board for the projects). Trevor Jahangard and then Chris Jenkins head off to a Hawaiian resort to work for a dolphin interaction program. Another student boards a plane to Bermuda to work with cetaceans there. Carissa Arellanes packs for far less tropical climes. She stuffs her suitcase with winter clothing and flies east to spend a week at the Cincinnati Zoo's Cat Ambassador Program.

On a Friday morning in early January, we set out across the broad parking lot at the Cincinnati Zoo—two trainers, Arellanes, I, and one serval on a leash. Mara, the serval, is so excited that the black line of fur down her back bristles. She bounces about the feet of her handler, Jennifer Good. The cat has tripped Good in the past with all her happy bounding. Mara flashes her tail back and forth. With the delicate steps of a ballet dancer in toe shoes, the serval's feet hardly touch the snow-dusted blacktop. Still, Mara leaves tiny little black paw prints on the field of white.

As we walk, the serval darts right, unreeling the black cord of her retractable leash. The cat lowers her nose to the snow, inhales deeply, then raises her small head topped with impossibly big, perked-up ears for a look around. Suddenly Mara throws her spindly front legs out and rushes back to Good in a couple of giddy strides. Good catches the serval between her legs and gives her narrow chest a rub. The cat's off again, this time to the left. She rubs up against Arellanes's legs, arching her back like a house cat. Mara's enthusiasm is infectious. None of us can help smiling, though the air is sharp with cold and a gray sky heavy with winter looms overhead. We sniffle, stuff our throbbing hands in our pockets, and walk into the frigid day behind this joyful African cat.

Mara resides in a small compound on a rise behind some trees to one side of the Cincinnati Zoo's parking lot. Here live two ocelots; a lynx; a young fishing cat; two cougars; four house cats; a male and female cheetah; and, surprisingly, one very large dog, an Anatolian shepherd. They are the zoo's Cat Ambassadors. These felines are trotted on leash into classrooms, where they strut their stuff. The ocelot scales an upright pole, then climbs back down head first. The serval, with its spindly front legs outstretched, hops high into the air. The cheetahs, well, just sit and be cheetahs, which seems to be more than enough to enthrall an audience.

The Cat Ambassador Program is emblematic of a seismic shift in thinking at zoos, the idea that captive animals can help save those in the wild. Many zoos these days have restyled themselves as conservation organizations. They invest in breeding programs to save endangered species, such as the Asian elephant, if only in captivity. They pair with conservation groups working in the wild. The Cincinnati Zoo has about thirty such affiliations, including the Cheetah Conservation Fund, Brazilian Ocelot Conservation Project, and the Fishing Cat Conservation Project. Moreover, zoos with their animals can give a faraway crisis some immediacy. If you see a mandrill in all his multicolored glory in a zoo, the thinking goes, you will be more sympathetic to its plight on the other side of the globe. You may even write a check. Thus the wordy panels hung by many a zoo cage detailing habitat destruction and the like, which can make a visit to the zoo a depressing meditation on human folly. The zoo's Cat Ambassador Program takes this approach one step further, by taking the animal out of the cage, out of the zoo, and into the everyday world. A cheetah sitting languidly on a desk leaves a much bigger impression than that same cheetah in a cage. "It's what we call the wow factor," Good says.

This program and others like it are made possible by trainers. Your average zookeeper does not know how to walk a wildcat on a leash. If you're going to bring exotic animals into classrooms, you better have someone who knows something about training on your zoo staff, and this is where EATM comes in. Three out of the four staffers for the Cat Ambassador Program are EATM graduates, including Good, who leads Mara back to her cage and collects Sihel, the ocelot. Sihel's cry is a fierce "REE-ow." She has chewed all of her toys. The ocelot is given a hunk of bone to gnaw on every night so she'll leave her own tail alone. Despite her wild ways, she is the first ocelot born from a frozen embryo; *sihel* is Mayan for "born again."

Compared to the prancing Mara, Sihel slinks out onto the blacktop, back hung low, eyes bright. She prowls the sides of the parking lot, where the snow cover is thicker. Good punts a chunk of snow her way. Sihel pounces on it. Another trainer heaves a snowball ahead of the cat. Sihel pees on it, then eyes a large, spiny stick. Good pauses so the ocelot can hoist it in her mouth. Sihel carries her head high, stretching her neck so she can drag the stick along. Before going back inside the compound, Good orders the ocelot to drop the stick for a treat. Sihel obliges, but a low grumble emanates from her chest.

Good leashes up Minnow, the program's new fishing cat, a young one that's not much bigger than a house cat. She has a pink nose and webbed paws for scooping prey out of the streams of Southeast Asia and India. Her gray coat is flecked with dark spots. We've just started down the sidewalk toward the parking lot when a long-limbed woman lopes up, takes a look at Minnow, and chortles, "That's not a cat. What is that?"

This is Cathryn Hilker. Hilker started what became the Cat Ambassador Program in 1981, when she began taking a cheetah into classrooms. Exotic cat trainers have a reputation for big, extroverted personalities, and Hilker fits the bill. In her seventies, she frosts her graying hair with blond. She wears big glasses that slightly distort her eyes. Her low, melodious voice naturally projects. There's something Lucille Ball about her, from her leggy tallness to her quick sense of humor. She relishes absurdity and tells stories that are funny at her own expense. When I ask how long she's worked at the zoo, she answers, "The only person who'd been here longer than I have is the hippopotamus, and he died." Hilker makes animal rights people froth, as you would expect, but she's rattled some zoo supporters as well. She once ran a cheetah on a race track. "Everyone was mad at me," Hilker says. "I thought it was so much fun. People are so uptight."

Hilker grew up on a farm in Mason, Ohio, a rural county to the north of Cincinnati, where she became an avid horsewoman; she showed; she foxhunted. She taught English at a private school, married at thirty-eight, and had one son. Then she adopted a ten-week-old cheetah. "Before cats, all horses and self-indulgence," she says. "After cats, my entire life is devoted to serving the cheetah."

Hilker's cause is the fastest land animal. A cheetah can hit sixty to seventy miles per hour in a few seconds, an acceleration any automaker would envy. Everything about the cheetah is designed for speed, from its oversized liver to its small head that offers little wind resistance. It's the only cat that doesn't have a collarbone.

Without it, the cheetah can stretch its front legs straight out for a stride that measures twenty-six feet long. The cheetah's flat tail acts like a rudder. The cat's pads are tough like tire treads.

The cat's speed, good looks, and relative docility have made it a favorite with man since antiquity. As early as 3000 BC, the Sumerians leashed cheetahs and hunted with them. Called coursing, hunting with cheetahs became an enduring sport from Europe to China. The cheetah wore a hood, as a falcon would, and once it was removed, the cat sprinted for the prey. Akbar the Great, an Indian mogul of the sixteenth century, reportedly had over 9,000 coursing cheetahs. The demand for hunting cheetahs drained wild populations, and by the early 1900s, India had to import cheetahs from Africa for hunting. Still, at the turn of the century an estimated 100,000 cheetahs roamed Asia and Africa. Now those numbers have dwindled to an estimated 12,500 cats, nearly all of which are in Africa.

The deck is hugely stacked against the cheetah, from the fur trade to habitat loss. African farmers consider them pests and regularly shoot them the way American ranchers once did the wolf. Added to that is the fact that today's wild cheetah population is deeply inbred, thanks to the Ice Age. That chilly stretch of time so reduced the number of cats that their gene pool today is a teeny one. Consequently, they are prone to disease and reproduce poorly. According to the Cheetah Conservation Fund, only one in ten cubs born in the wild makes it to adulthood. Though zoos have had luck breeding many species, the cheetah is not one of them. That this specialized marvel is still with us on planet Earth is no small miracle.

However, this is not what Hilker was thinking about when she first proposed to the director of the Cincinnati Zoo getting an animal or two out of their cages. As she remembers, "He said, 'That's the dumbest idea I ever heard.' " In the seventies, her idea was revolutionary. Most zoos then were museums, where animals were displayed like living sculptures. Hilker got the okay to get a horned owl out of its cage. She wasn't to do anything but hold the bird and talk about it. Whenever she did, visitors surged around her. "When you get out behind the bars and out behind the moat, you give the animal an immediacy," Hilker says.

Hilker asked a friend, Jack Maier, then president of Frisch's Big Boy restaurants, to donate $5,000 to start a program of taking a few animals—a snake, the horned owl, an opossum—to schools in what became the zoo's outreach program. Then she got Maier to fund Frisch's Discovery Center, where the public could see some animals up close, even touch some. She asked the zoo director if one of the lion cubs that was rejected by its mother could move to the center. He said, "Yes, but don't take it off the grounds." Hilker took the lion cub home at night and brought it to the zoo each day. People lined up ten-deep to see him.

She was on a roll. She asked for her very favorite animal, a cheetah. "To my utter astonishment, I was told I could go to Columbus and get one." The Columbus Zoo loaned the Cincinnati Zoo one of two cubs. The cheetah was about the size of a house cat and still had her mantel, the long, fawn-colored fuzz that grows along a cheetah cub's back. Her name was *Maliki,* Swahili for "Angel." "I had no idea what it was like to have a wild animal give you its heart," she says. "Everything about it cut me like a knife."

For the first year, Angel slept with Hilker and her husband. She played with the couple's harlequin Great Dane, Dominic. By the time she was a year and a half, Angel was a fully grown, eighty-pound cheetah and stood twenty-eight inches. She measured five and a half feet from her nose to the tip of her tail. While still a cub, Hilker began taking the cat to schools. The cheetah was called the zoo's Wildlife Ambassador of Goodwill. Angel sat and Hilker talked. She talked about cheetahs, about wild animals, why they are important, about the "web of life."

Zoo honchos were worried they would be criticized for having a cheetah on a leash, Hilker says, but the public and the media were fascinated. "Money started to come in." The twosome became national—even international—celebrities of a sort. They made the rounds of national talk shows. Angel licked David Letterman and jumped on the couch with Regis and Kathie Lee. They met Prince Charles in Palm Springs.

Zookeepers in Cincinnati begrudged her her success, Hilker says. Angel, they sniffed, was not a *true* wildcat; she'd been tainted by training. Training was a bad word at zoos. The T word has circus connotations and bespeaks tricks, even costumes. Zoos considered themselves serious educational institutions and anything that verged on entertainment was suspect. A cheetah on a leash sounded like too much fun. "It was totally disapproved of," Hilker says. Still, she says, "Some of the zookeepers who were the most vocal critics would call and say, 'Hi, my parents are in town. Could they come see Angel?' That was very rewarding."

Angel was so in demand, the Cincinnati Zoo enlisted more cats. Carrie, a mountain lion Hilker raised on her farm, "a teeny, runty cat," she says, was drafted into service. Then the serval Mara and a snow leopard moved in. The Cat Ambassadors became one of the zoo's most visible programs and a national model. "Now everyone wants to mimic the program," Hilker says. "There have been some horrific ones because they don't want to invest in the training. Cheetahs are not pussycats. When they get angry they strike."

You just don't hook up an exotic cat and go for a stroll as you would a dog, though trainers who know how to walk a tiger or cougar make it look that easy. Mara Rodriguez does. At EATM, Rodriguez walks at least one of the two cougar brothers, Sage and Spirit, nearly every day she's there, chaining a cat and hooking her fingers with her long manicured nails through the links for a better grip.

"I seem calm and mellow [when I walk them], but if people could only see inside my head with everything going on and all the gray hairs under my dye job," she says.

Rodriguez has been walking big cats for twelve years. When Rodriguez was a student at EATM, none of the cats came out on a leash, so she learned on the job at Animal Actors of Hollywood. Rodriguez started with cubs, which weren't as scary, and worked her way up to tigers. Along the way, she built up the needed confidence. "Working with cats, you don't even get okay until about two years," she says. "You have to walk without fear. . . . Once you look comfortable to other people you're getting somewhere."

When Hilker decided to leash up big cats, she knew she needed help. She had trained dogs and horses but never a wild animal. She hired two trainers to coach her: first, a woman who had leash-trained a cheetah at Six Flags Marine World; second, a sarcastic EATM grad, who spent three summers teaching Hilker. "She was merciless all day, every day," Hilker says. "I'd go home and cry on the way home. I paid her my money. Me. My money. And she was merciless."

To walk a big cat is to learn how to bluff. You want them to think that you are in charge, though the cats are stronger, faster, and have sharper teeth. The bluff comes from being confident and consistent. Hilker learned to not let the cat ever get away with anything, and to always know in advance what she wanted the animal to do and where she wanted to lead it. She learned how to get a lounging cat up for a walk. She learned that different cats require different approaches. Hilker needed to be forceful with Carrie, the cougar, who preferred sitting to walking and was bullheaded compared to the cooperative Angel. Cougars don't take offense at being corrected, but if you discipline a cheetah, Hilker says, "they are in an instant snit. They will close their eyes and go rigid. . . . They are Miss Priss with high heels and a tutu on."

Angel was a perfect starter cat for Hilker. She was forgiving of the trainer's early blunders. In eleven years, Angel struck Hilker only once. During a presentation, a kid crawled under the table Angel was sitting on and pulled the cheetah's tail. Spooked, Angel swatted Hilker and her dewclaw caught the trainer's arm. The cheetah watched in fascination as blood ran down Hilker's arm.

Angel died in 1992. In her memory Hilker created the Angel Fund. Donations to the fund support the Cheetah Conservation Fund, which works to save the cat in Namibia. Namibia has the largest population of wild cheetahs, but their numbers are dropping. Using an innovative approach, the program gives farmers Anatolian shepherds, an ancient Turkish breed, to protect their livestock. With the dogs protecting their livestock, the farmers don't feel compelled to kill cheetahs.

This explains why there is a dog in the Cat Ambassador Program. Alexa joined the program as a puppy so that people could see the breed that was helping save the cheetah. While in Namibia the shepherd and the cheetah are not on friendly terms, here in Cincinnati they are boon companions. Sara, one of the program's two current cheetahs, and Alexa got to know each other out at Hilker's farm. At first there was plenty of fighting, but now the twosome are very simpatico. At the zoo, they are roommates. They still play, but Sara tires of it first, whereas Alexa could wrestle all day. They are only separated at feeding time. Dinner can bring back the old growling.

Though Hilker is quite underdressed for the winter day in a sweater and silk scarf, she wants to show me how you walk a cougar. It's near closing time, 5 P.M., and the pale winter light has just begun to fade. Hilker grabs a heavy chain and heads into one of the cougars' outdoor enclosures. From where I'm standing,

I can't see her but I can hear her. "There you are, you big ugly cougar," she says. "Why are you sitting there looking at me like you might attack me? Get over here, you mean thing." She emerges with the fawn-colored cougar, holding the chain in her gloveless hand. Arellanes and Good join us. The cold clamps on us like a vise as we trod across the parking lot for the last cat walk of the day.

The cougar is matter-of-fact. She neither bounds like the serval nor hunts like the ocelot. "She just trundles along," Hilker says, kicking some snow the cat's way. Hilker hears a car and stops. Cars can rattle the cat. When the engine's purr fades into the distance, we turn to go. The cougar remains seated. This is what Hilker wanted to show me. She applies steady pressure on the chain, and before long the cougar obliges. A cheetah treated this way would pitch a tantrum, Hilker says.

"Let's leave cougar prints," she says. "I love leaving cougar prints."

Just outside the parking lot's fence, brick houses line the street. The low winter sun glints pink in their windows. We cross the snow-glazed lot, a trail of paw prints behind us. We arrive at two picnic tables. Hilker orders the cougar up on one, then tells her to leap to the other, which the cat does in a flash. When Hilker asks the cougar to sit, the cat doesn't budge. Good tells Hilker she's using the wrong hand signal. "Oh," Hilker says; she points her index finger at the sky, and the cougar complies. Arellanes and I pose for photos. We pull off our gloves and dig our fingers into the cougar's rough, dense fur, like a coarse rug. We smile for the camera. The cougar purrs.

Trainers have a great and unusual gift to share, to make the impossible possible, and most are generous with it. Hilker is no different. After the cougar is put away and the staff has left, she turns to me and asks, "Want to pet a cheetah?" Before I know it, we are standing outside Sara's small indoor cage as Hilker looks me over for anything that might catch the cat's eye. She has me put down my bag and take off my gloves. She wonders out loud about my coat with its loose, nubby weave, but rules that I can keep it on. A small distant voice in my head sounds an alarm, but when Hilker opens the cage door I follow her in.

Sara lounges on her raised, rectangular bed filled with hay. Hilker sinks down next to her and sighs happily. This is the first she's seen the cheetah since a recent trip. "I couldn't wait to get back and thump my cats," she says. Hilker drapes her right arm over Sara and scratches her neck with her left hand.

I do not exude confidence, and I know it. I'm less scared than awkward. I feel like I should curtsy or bow. Hilker motions me to sit down on a second bed, close to Sara's, and to pet her on her shoulder. I reach out and gently touch her. The cheetah doesn't even turn her head. Her shoulder is bony. Her fur is like chenille. I'm dumbstruck by the sight of my own hand against cheetah spots.

I tentatively stroke Sara with my fingertips as Hilker chatters casually about her vacation. I can hardly respond at first. A satisfied purr sounds deep inside the cheetah. The cat turns to look at me once or twice but mostly can't be bothered. Before long, I find myself ruffling her fur with my fingers as if she's a dog, and talking away to Hilker. Suddenly, two women chatting in a cage with a cheetah between them seems the most natural thing in the world, and that's the marvel of it.

That evening, Hilker leans against a table with Sara posed on one side and Alexa the shepherd on the other side. About a dozen people have turned out on this frigid night ostensibly to hear Hilker speak, but really to be in the same room with a cheetah. The small group smile daftly and sit up straight in their seats. "All my friends were jealous when I told them what we were doing tonight," a woman gushes to Hilker. Hilker leans against a table, her long legs crossed at the ankles before her. "You're a good kitty," Hilker says. For the umpteenth time in the past twenty-five years, Hilker tells of her true love, this cat with the flexible spine and aerodynamic tail. She points out Sara's dewclaw and the black teardrops under her eyes. The cat and dog hold their heads high and stare straight ahead while two Cat Ambassador staffers clutch their leashes. As if bored, Sara raises a paw and licks it gently. "Are you primping?" Hilker turns and asks. The cheetah shifts, as if to get down, and a trainer stops her.

On the other side, eager-eyed Alexa pants. When she's asked to speak on command, Alexa woofs enthusiastically. When the trainer leads the dog through the small audience, Alexa slaps the air with her tail and solicits pats with her bright eyes and soft, moist nose. In Namibia, one of these dogs killed two male baboons, Hilker says.

Then it's Sara's turn to step down. The cat refuses. She's peeved about being stopped earlier. It is, as Hilker says, "so cheetah." After much coaxing, the cheetah hops down as if it were her idea to begin with and saunters back and forth through the audience, her shoulder blades shifting dramatically from side to side. She slinks like a runway model. People lean forward in their seats, crane their heads for a better look, and sigh. A man at the end of a row bends at the waist so that his face is even with Sara's as the cheetah slides by. The wow factor is clearly at work. Hilker smiles yet again at this simple magic. "Isn't it charming to have a cheetah walk right by your face," she says. "Aren't we silly?"

Falling in Love

Love is in the air at the teaching zoo these days. The first years ponder their affections for various animals, divining crushes from true devotion, considering compatibility versus sheer magnetism. One enthusiastically proclaims her love for the turkey vulture to me. Another confides he's got a thing for the binturong. A first year from Virginia has given her heart to a nippy macaw. All this lovesickness has a reason. The first years must turn in their animal requests in less than a week, by January 21. These will be the animals they will work with come summer semester, when the second years have graduated and gone. Summer is still four months away, but it's the first light on the dark horizon, the first promise of meaningful animal contact to come.

The first years must pick animals in four categories: primates, carnivores, birds, and hoofstock. In each category they can list their choices in descending order, from most desired to least. It's something like making a Christmas list. And who gets what will have something to do with who's been naughty or nice. These assignments are the big payoff for the months of hard work.

In the Briz years, he pretty much chose who got assigned to what animal. Now there is a system with a scientific bent. The staff ranks the students according to grades, attendance, and how much time they have weeded or painted around the teaching zoo, what they call volunteer hours. The students with all the possible points, such as Linda Castaneda, Susan Patch, and Terri Fidone, land at the top of the list. The animals in demand, such as Schmoo and Taj, go to these students. Then the staff starts working down the names. The lower your grades, the spottier your attendance, the farther down your name falls. If you are near the bottom, well, you'll get the animals that are left, the ones no one wanted.

This system is the reason why that one morning Larissa Comb rolled in thirty seconds late haunts the former stockbroker. She has straight A's, worked all her volunteer hours, and was only late that once. But that one blunder sent her name trailing down the list. Still, Comb takes her chances and asks for Schmoo, though she doesn't have all her points and knows about a dozen students would like the sea lion.

Tony Capovilla does not have straight A's but he can pretty much expect to get the only animal he wants: Rosie. The former plumber came to EATM just to work with her, which was a bit of a risk. There are four men among the first years. As it turned out, though, one, a former Federal Express driver and father of two, does not want to work the baboons. He thinks they are scary, not to mention time consuming. He'd rather put the hours into Max, the neurotic military macaw he's fallen for. So that nearly guarantees Capovilla Rosie.

These requests are especially freighted, because this is when the first years choose what they call their year-longs, which is what Rosie is. Most of the animals at the teaching zoo get a new set of trainers every semester, but some the students work with for twelve months. This includes all the primates and Schmoo. A year is quite a long engagement, so the students gnash their teeth, scheme, and strategize.

Anita Wischhusen tells me disgustedly that some first years keep their animal requests to themselves. "They say, 'I'm not going to tell you because you'll ask for them.' That so doesn't make sense." Wischhusen does not have the grades for Schmoo, so she won't ask for her. She doesn't quite for Taj, but she's requested the tigress anyway. In the bird category she's asking for a "turaco, turaco, and turaco. I don't want a bird who will bite. I'm not a bird person." She's leaning toward Zulu for her primate.

Patch is one of those first years keeping her choices to herself. She's trying to avoid the competition, jealousy, and resentment that swirls around animal assignments, but, as Wischhusen makes clear, that's nearly impossible. Patch's first picks are Goblin, the hamadryas baboon; Laramie, the eagle; Sage and Spirit, the cougar brothers; and Lulu, the recently arrived pregnant camel. Like Wischhusen, Castaneda does not want a parrot. Like Patch, she wants the cougar brothers. She'll also ask for Scooter, the capuchin, and the two-for-one deal, Walter and Dunny. She figures the young water buffalo and the Scottish cow will be a good way to build her confidence for a camel later.

First years also consider who will be their co-trainers. As many as four students are assigned to each animal. So if you hear a fellow student—a known slacker or malcontent—is asking for the same animal as you, you might think again. There are a first year or two who fall into this category. One in particular has a reputation for worming out of work. Bets are on her, a young student with unkempt long hair and moist eyes, being the next to wash out.

As with all things at EATM, the animal requests provoke much discussion and anxiety. Things here so naturally rise to a fevered pitch, if only because everyone is so exhausted. The first years fret like brides waiting for word of their arranged marriages.

In their more levelheaded moments, the first years say they will love whatever animal they get. Time after time that has proved true, says Dr. Peddie.

For nearly every animal, there is some wide-eyed, love-struck student. "I never thought I'd fall in love with this cavy," Mary VanHollebeke says. Her affections crept up on her even though, as she says, the cavy "doesn't give much back." When she thinks of her spring graduation, she says, "He's the animal I'll miss the most." Kage adores Benny, the ancient capuchin, even though he's bit her, masturbates regularly, and resembles a character from a Stephen King novel with his hairless, curled tail. Two second years are devoted to Louie, the paraplegic prairie dog that rarely emerges from his den box. Even the teaching zoo's gardener fell hard.

The tall, muscular fellow with a constant tan and ready smile dotes on Sequoia, the mule deer. One day as he was nervously trimming the acacia bush by Sequoia's cage, the deer trotted over. She poked her soft muzzle through the fence and licked his arms. At first, he worried that Sequoia might bite him, but before long he was smitten. Now he brings the deer bundles of browse to chew. When the east winds gust, it rattles Sequoia, so he sits outside her cage to calm her. Last year, when Sequoia pummeled a second year, the gardener found himself terribly torn. He was dating the student. "I thought, poor Stephanie. Then I'd think, poor Sequoia." Eyewitnesses claim he yelled, "Don't hurt her," meaning Sequoia.

Becki Brunelli has developed a big soft spot for Nuez, the Central American agouti. "I like rats so much and he's a big rat," she says. "He's kind of an underdog and I always like underdogs." He's the size of a house cat and has a shiny cinnamon-colored coat stippled with black. His head is shaped like a squirrel's but with a longer muzzle. At first glance he looks tailless but between his rounded haunches is a tiny bald nub, just enough to remind you he's a rodent.

Brunelli is one of the few, if not the only, students who will pick up Nuez and cuddle. That's because Nuez is known as the rodent squirt gun. Male agoutis spray female agoutis with urine as a come-on of sorts. Nuez tries the same technique on the students. Just inside the cage door Brunelli keeps a plastic cafeteria tray to shield herself from Nuez's love sprays.

Many of the EATM students dislike rats to begin with, especially one that squirts. A number of the students assume it's ejaculate that Nuez jets at them. One calls it his "special sauce." Brunelli always corrects them. It's urine, she reassures them. Brunelli has lobbied hard for Nuez, urging first years to request him as an animal assignment. "Whenever anyone says anything bad about him, I always say, 'But he's so cute.' "

To prove her point to me, we head down to Nuez's cage behind Nutrition. Brunelli swings open the door, and Nuez rises up on his back legs. "Don't you spray me," Brunelli says. She emerges cuddling the agouti like a baby, his soft nose tucked into her neck. He rubs his face along her throat, and gives her a peck on the cheek. He's so happy he squeaks. Two first years wander over, curious.

"Is he a rodent?" one asks.

"Does he bite?" queries the other.

Just nibbles, Brunelli explains, love nibbles. She launches into her pitch about how affectionate Nuez is while the rodent snuggles close. First years often look on with obvious envy as second years embrace their animals, but not now. The two first years look dubious, even a bit revolted. "He's trying to seduce you," one says to Brunelli.

"They mate for life," Brunelli says, which I don't think is much of a selling point—at least not for these two.

Brunelli isn't the only second year actively lobbying first years, urging them to ask for their animals. Kage has recruited a first year for Todd, the fox, and has groomed another for Benny. Van-Hollebeke pleads with first years on her cavy's behalf. Oddly enough, Amy Mohelnitzky finds herself lobbying for Abbey. She loves the dog so much that's she asked to train her again this semester, but none of the first years are interested in the canine star. To them, she's just a dog—moreover, a high maintenance one that needs bathing and brushing and fussing over her food. Why would they work with a dog that has a bad stomach when they could train a perfectly healthy lioness?

It doesn't help that Abbey has been more miserable of late. Mohelnitzky picks a staffer's brain on what might calm the poor dog's stomach. Nothing has worked so far. As they talk, the staffer mentions that Abbey would do better in a home. Mohelnitzky raises her eyebrows. "Would you want her?" the staffer asks. Not long after, Mohelnitzky asks her apartment manager if she could have a dog. He says no.

It's the first week of second semester. A few of the big carns, such as Legend, have leftover Christmas trees in their cages. They can pounce on them as if they were big prey. The lemon trees hang heavy with fruit by the pen of Clarence, the Galapagos tortoise. The mornings dip below freezing, and the sun is slow to emerge over the mountaintops to the east. The students don headlamps like coal miners, aiming the beams through the dark at cage floors, scanning for wayward turds.

The first years gird themselves for what they are warned will be an even tougher semester. How can that be? they wonder. This is how: in addition to classes, the first years will be the worker bees for the Spring Spectacular, the school's annual fund-raiser. That means cleaning the zoo, studying for class, building sets and props, and tons more schlepping in general. A student from Argentina gets to shaking just thinking of the semester ahead. Susan Patch, as usual, remains unflappable. She's signed up for kick boxing this semester.

The first years' reputation sinks a little lower. A classmate went in with the capuchin troupe by herself to clean, which is against the rules. She thought she had closed the chattering monkeys off to one side of their baleen cage, but the alpha capuchin slipped through to her side. She fled, leaving tools behind. Another first year opened a door in Big Carns too soon while moving an animal into the arena. Castaneda and the other go-getters among the class resent being lumped in with the screwups. How will they ever measure up to the second years at this rate?

For their last semester at EATM, the second years have relatively few courses remaining. The focus of all their energies will be Spring Spec, specifically "Welcome to the Jungle," a kind of hallucinatory, wordy tale of a circus train crashing on an island. The show has a big cast and includes many cameo appeareances by teaching zoo animals. By March, the second years must learn their lines and teach the animal actors their parts.

Before any training can start, though, the turnovers have to be complete. This process always riles up some of the animals. Samburu the caracal has gotten so aggressive during feedings, knocking his face into the bars, that he's skinned his nose. Salsa, a Catalina macaw, has gotten nippy. Cain, the chattering lory, wouldn't quit biting Brunelli, so she quit handling the bird. Instead, she sits outside the cage, hoping to make friends from a distance.

Depending on the animals, turnover can take hours, days, or even months. The process for Birdman, the kinkajou, is simple. Birdman's current trainer demonstrates how to crate the critter, and Jahangard's set to go. While Jahangard is in the cage, though, something funny happens. The kinkajou opens his mouth wide as if to bite the student. His current trainer, who nuzzles the kinkajou with her face, has never seen Birdman do that. On the other end of the spectrum, the turnover for the cougar boys can last the whole semester. If all goes according to plan, by that time each of the students will have walked a cougar on a chain around the teaching zoo.

Mara Rodriguez runs the changeover. In fact, the students never work with the cougar brothers without Rodriguez present. This afternoon, she steps into the arena in the center of Big Carns with Sage and Spirit. As usual, Rodriguez is put together with her well-fitting clothes, mirrored wraparound sunglasses, and long nails painted a bruised maroon. In a corner of the arena, two second years keep Spirit busy so Rodriguez can work with Sage, the mellower of the two cats, on the opposite side.

Rodriguez starts each turnover by acquainting the students with the cougars and vice versa. This is a case where desensitization works both ways. However, the brothers have a head start. Sage and Spirit are far more used to humans than these three young women are accustomed to standing next to a mountain lion. To work with the cougars, the students have to learn a delicate balancing act: how to be relaxed yet attentive, confident yet cautious. They have to find that spot where opposite impulses harmonize. If they don't, they risk getting hurt or worse.

In the afternoon sun, three second years lean against the high chain-link fence, staring, as if hypnotized by Sage. The cougar, his eyes closed, lounges on the ground by Rodriguez's feet. Rodriguez calls in one of the second years, a pear-shaped young woman with straight hair down her back. She steps through the gate looking less frightened than awkward. Rodriguez tells her to tuck in the dangling strings of her sweatshirt hood and then motions her over to Sage. "You can pet him," Rodriguez says. "You can lean down to do so but don't ever reach down and tie your shoe or massage your ankle."

The student squats and tentatively strokes the cat. Sage licks a round paw and scratches his head. He never looks at the second year. Off in the corner, the students playing with Spirit laugh and gossip.

"I'm going to have you walk from here to here, and don't look weird," Rodriguez tells the student. "I don't want you looking around, running into things. He'll notice that. Be confident." The second year walks a few steps one direction and then back as we all watch. She looks self-conscious. The humans watch her. The cat does not.

That student comes out and the next goes in—Jena Anderson, the second year who taught Nick to pull the cart. Rodriguez has her tuck in her sweatshirt strings also. Anderson, who typically is poised, now looks just the slightest bit bug-eyed. Spirit wanders over from the far corner. Anderson turns, sees Spirit, and freezes. The second years in the corner of the arena call the stray cougar back. "You just get the hell out of the way if [the cougars] look like they will chase each other," Rodriguez says.

Sage remains detached. He pulls himself up and saunters in Anderson's direction. Again, Anderson freezes. "If they don't pay attention to you, it's no big deal," Rodriguez says, walking alongside the cat. Then, as Rodriguez steps over a log lying on the ground, she catches her pants leg on a branch. She almost trips—exactly what you don't want to do. Rodriguez catches herself, stumbling ever so slightly.

Anderson switches off with the third student, who stands around the drowsy cat. She walks from here to there, as Rodriguez tells her, tries not to act weird, and tries not to act like she's in a cage with a dangerous animal. The indifferent cat stretches out on the ground, lazily squinting. "It seems boring, but I hope you guys appreciate that there are people in the cage," Rodriguez says. "You want [the cougars] to not care."

* * *

Come second semester, first years can sign up for the first of three classes on primates taught by Cindy Wilson, Gary's wife. The couple began dating in high school and went to EATM together. She works part time at the zoo. In addition to the primate classes, Cindy oversees the behavioral enrichment program. She is prone to small, wry, almost inscrutable smiles and is typically soft-spoken, that is, until she uses her primate voice. Then she booms, so most any primate, including a human, will stop in its tracks. She's a motherly figure to many of the students, a role she enjoys. A past class nicknamed her Momma Cindy. She's been known to hand out Twinkies to students who answer questions correctly in class.

Cindy Wilson learned about primates from working with them firsthand. When Gary was the director, she stopped by the college to pick up his paycheck and came home with a six-day-old rhesus macaque. "At first I was overwhelmed," she tells me. "Within twenty-four hours we had bonded. I said, 'He can stay.'" His last feeding was at midnight. By four months, he could scale the refrigerator. Gary built a jungle gym of PVC pipes in their living room. Cindy had found her niche. "With primates I'll never know all I need to know," she says. "It's a very dynamic relationship."

She's using today's class as an introduction to some of the primates, knowing the first years have to finalize their animal requests. So we walk down the front road en masse to Zulu's cage, where several of his second year trainers stand waiting. Zulu sits on his shelf behind them, nonchalantly chewing gum. Kristina Nelson, the coolheaded Iowan whom Schmoo charged last month, starts. "A lot of people pass on asking for him because you have to deal with his owner," she says. "Everyone who comes with Zulu is very nice." That pretty much concludes the sales job. What follows sounds like a support group for abused wives, all married to the same polygamist husband.

"He's spoiled rotten," Nelson continues. "He's the biggest spoiled baby. He's left bruises on us."

"It's hard because he threatens the whole time," chimes in another trainer, a reedy brunette. "He grabbed me a lot. I quit grooming with him in the summer." While she talks, Mary VanHollebeke blows slowly into Zulu's mouth to keep his highness happy.

"The first time he grabbed me, my legs shook," Nelson says. "You have to pet him, not groom him. You have to go the right direction on his fur." If you don't, he'll grab you, she adds.

Someone nearby mutters, "I don't think so." Everyone squints into the sun and looks dubious. Cindy explains about the harem Zulu would have in the wild and how he has to assert his authority over his female trainer. The first years look at her, nodding blankly.

"He scared me more than anything," says the brunette. "I have scars on my hands from his nails."

"Even feeding at first was terrifying," says Nelson. He's left bruises in the shape of handprints on their arms. You have to let him squeeze your arm, she insists, or he'll stay mad. Maybe sensing that she's overdone it, Nelson adds, "When you go away and come back, he's so ecstatic to see you."

Zulu, one arm draped over a knee, looks as if he's calmly listening, carefully considering all that's being said. VanHollebeke softly strokes the side of his big head, which he leans up against the bars. As the group shifts around the corner to Goblin's cage, there are some shared glances, some mutterings about not wanting to be grabbed. Sunni Robertson, who took care of Zulu in the fall, has decided the mandrill is not for her. Wischhusen, who's afraid of the macaws, is unintimidated. She will ask for the mandrill, her "monkey in drag."

A second year with big brown eyes leans her back against Goblin's cage. The baboon carefully works her black fingers along the second year's hairline, careful to leave the student's barrette in place. Cindy explains that the second year grooms with Goblin so the baboon will feel comfortable with the group of students gathered round. It's a show of solidarity, the second year's way of saying to the baboon, "I'm on your side." Goblin loudly smacks her lips together, a sign of contentment and friendship. The only downside to Goblin, the brown-eyed second year says, is that her turnover will be very slow, but otherwise she is a honey. The first years smile. Some even sigh enviously. This is more like it. Who would volunteer to be grabbed when they could be groomed? The first years are, after all, only human.

The next day, Saturday, a few second years dawdle in the chill morning air by the wolf's cage. Legend bounds about excitedly, her white tail flashing, her lips parted. Even C.J., the coyote, is wound up in the cage next door. The coyote bites at her hind leg, turning tight circles as she does. Holly Tumas strides up and pulls out a heavy key chain. Tumas, who has taught at the school since 1999, has her baseball cap pulled low over her long, straight, dark brown hair. Former male students nicknamed her "Hottie." She laughs a great joyful laugh. Though she is a stickler for the teaching zoo's rules, she is approachable and friendly. She is Legend's alpha female.

To be Legend's alpha takes more subtlety than you'd assume. You can't throw your weight aound; you'll scare the wolf. Scared, Legend becomes aggressive. With her upright posture and solid voice, Tumas broadcasts that she is top rank. Her long friendship with the wolf helps. Tumas has known Legend for six years, since she graduated in 1997. She got a job with the animal show at Magic Mountain, a theme park in Valencia, where she trained and walked Legend, or Ledgy, as she is called at EATM. When the park sold off its animals, Tumas and then, later, Legend joined

EATM. If the wolf ever second-guesses Tumas's dominance, she takes over a few of Legend's feedings in order to reestablish that she is top wolf in this small pack.

Female students typically need speech lessons to work with Legend. As one second year says, she's working on her "Legend voice." The natural female speaking pattern—that musical jingle—does not lend itself to training a wolf. The students can't use the high, chirpy singsongs they so often do with the other animals. They can't slide up a note at the end of a command, making it sound like a question. They have to make their voices lower and firmer. Men, however, must do the reverse. Men, especially the more testosterone-laden, make Legend nervous. Even those with a soft touch, such as Chris Jenkins, can unnerve the female wolf. Consequently, male students learn to tone down their masculinity.

Legend calls a corner cage along the front road home. Her archenemy, Kiara, the lioness, is her neighbor on one side; her best buddy, C.J., the coyote, is on the other. The wolf's coat is the color of oatmeal, except for the darker patches that outline the bridge of her nose, the points of her ears, and the curve of her haunches. She has amber eyes and a jet black nose. Tumas says she is an extra-social wolf, a trait that has deepened as Legend has aged. Now ten, she craves the student's company. "She's such a good wolf," Tumas says. "It's common for people not to think of her as a full wolf. It's so wrong. Her aggression can come lightning fast. I've seen it." Legend also has a thing for Michael Jackson's "Man in the Mirror." Whenever she hears the pop ballad, she howls.

Somehow Legend always knows when she's going for a walk. That is why she and C.J. are excited. Though the coyote is not coming out, the two canines consider themselves members of the same pack, so what happens to one has an effect on the other. That Legend is a pack animal makes her training altogether different from the cougars'. The cats are solitary creatures and don't require social attachments. Legend, however, yearns to be in a group. To train and walk the wolf, the students must become her pack members. This is a process dictated by the wolf. Some students Legend welcomes into the pack faster than others. She readily accepted one of the second years because the student trains C.J. If the coyote counts her as a pack member, the wolf automatically does too.

The key to gaining entrance to Legend's pack, Tumas says, is to take it slow. "If [the students] push too hard, it sets them back." As a first step, you sit near the wolf's cage. Eventually, you talk to Legend and feed her. Then you wait for Legend to sniff your hands or offer her furry flank along the enclosure—what's called presenting. Once you have that stamp of approval, you step into the wolf's cage with Tumas. There, you stand and wait for Legend's next signal. This may take several sessions. Once Legend rubs her legs against you or smells your hand, you are an official pack member. Only then can you learn to walk the wolf.

Taking Legend out of her cage is something like the changing of the guard. It's a highly regimented system during which Tumas announces every move, such as the lock being unlocked, and the cage door being opened. Tumas steps through the door first, closing it behind her. Then a student does the same. When they are in the cage, they never, ever turn their backs on Legend. Meanwhile, as Tumas commands, Legend sits on a tree stump in the middle of the cage. Today, Legend is so excited, like a dog anticipating a trip to the park, that she skips off her stump and prances over to Tumas. "On your mark," Tumas orders in her wolf voice. Legend does as told by her alpha.

A slouchy second year, who has recently become a pack member, squeezes in behind Tumas. Today, she will learn the first step of walking the wolf, catching her up. The two women move to either side of the wolf. The student holds out a looped rope. On her command, Legend leaps forward off the stump as the student kind of lassos her. Her aim is slightly off, but the wolf's head goes through the loop, if just barely. While the student holds the rope, Tumas leans over Legend and hooks on a heavy chain.

Legend passes through a series of doors, and the band parades up the back road. Tumas grips the chain today, but eventually all the students will take it. For now, they take up assigned positions, like sentries, around the wolf. Everyone hustles to keep up. The cougars poke along on their walks, meander from here to there. Legend is all business. After stopping to pee on a eucalyptus bush, Legend strides along in a straight line, hardly breaking her pace to inhale a smell here and there. Her head held low, she pants lightly. The chain jingles. Legend pauses briefly to press her nose to a building. "Stopping!" a student out front calls. Then we're off again.

Tumas keeps an eye peeled for squirrels and rabbits, even a dog being walked outside the zoo's fence—anything that might catch Legend's predator eye. If the wolf was to sprint, Tumas would dig her heels in, cling to the chain, and lean back like a water skier. Tumas has never had to do that, though she flexed her biceps hard once when Legend lunged at an empty animal crate.

We turn into Wildlife Theater, which is empty just now, though visitors have begun to trickle in to the teaching zoo. Tumas orders Legend onto a tree stump in the middle of the earthen stage surrounded by empty metal bleachers. The wolf neatly steps up and balances gracefully atop the log. A little boy wanders around the bleachers and stops dumbstruck. "Is that a wolf?" he asks.

Legend's eyes remain on Tumas. "Good stay," she says. Tumas and the students raise their chins and begin to howl. Legend looks from person to person, cocking her head curiously, and flutters her tail. The humans persist. Legend seems nearly to shrug her shoulders, then raises her black nose and pushes her ears back.

She joins in with a rich howl heavy with beauty and longing. Legend's call resonates in your chest, fills you up. Tumas and the students hardly touch the real thing. They let Legend take the lead, and the small choir raise their voices to the heavens above like countless packs before them.

Birdman Bites

During a morning cleaning in January, Trevor Jahangard strides the length of the zoo and pushes the chain-link gate to Quarantine open. This is where Birdman, the kinkajou, lives. Jahangard's low-key manner belies how much he accomplishes every day. In addition to EATM, he's apprenticing to be a falconer and taking night classes in calculus and chemistry this semester. Like many EATM students, Jahangard discovered his love for animals early on, but in an atypical way. When Jahangard was six his family went to visit his father's relatives in Iran. At nearly every house they visited, the host would ritualistically kill a sheep, slit its throat, say a prayer in Farsi, and then butcher it in the backyard. "I was fascinated. I told my mother I wanted to be a butcher. I told her I wanted to cut up animals. She said, 'Why not become a vet?' "

Neither of his parents cared much for animals. His mother is allergic to rabbits, cats, and dogs. His father and his side of the family revile snakes and aren't in the custom of keeping pets. Still, Jahangard's parents let him fill terrariums with snakes, salamanders, and frogs. At nine, Jahangard bred and sold lovebirds. He bred cockatoos as well. He attended a magnet school at the Los Angeles Zoo and from there came directly to EATM. He wants to train primates.

Jahangard asked for the kinkajou this semester because he wanted an animal that wouldn't take too much of his time. He's busy enough training Rosie and working on Spring Spec. Still, like most EATM students, he has big ideas about what he can accomplish in a few months' training. He plans to train the kinkajou to come out of his cage on a leash. This will be for a grade. The fact that the kinkajou locked on to VanHollebeke's hand last May does not worry him. He's trained Kaleb, Rosie, and the cougar brothers. A small member of the raccoon family weighing only five or so pounds shouldn't be a problem.

Kinkajous are enchanting, otherworldly-looking animals, with their wide shiny eyes, little pink snout and ears, long prehensile tail, and narrow, serpentine tongue. The underside of their paws, crossed by lifelines, looks shockingly human. Their feet rotate 180 degrees. Their teeth are as sharp as a predator's though they prefer fruit, insects, and an occasional long drink of nectar. Their coats are waterproof like a beaver's or otter's.

They snooze away the day high above the ground in the rain forest canopy of Central and South America. They are often called honey bears and kept as pets. The word around EATM is that Birdman formerly belonged to the actress Kirstie Alley. Now the kinkajou bunks in a cage next to Wakwa, a raven. Here he sleeps curled up in a blanket in his den box. He does not rouse easily. Students can spend as much as a half hour waking the kinkajou for a training session.

Jahangard's been told Birdman has diarrhea. A first year who may ask for the kinkajou for this summer (the Georgian who cared for a houseful of baby deer) tags along with Jahangard. Birdman does have the runs. The kinkajou has also pulled his blankets out of his den box and, uncharacteristically, he is on the cage floor. Jahangard goes in to put the blankets back, but he's hardly stepped inside when Birdman springs at him and latches on to his knee. Jahangard quickly knocks the animal off and sets Birdman down on the cage floor. Before Jahangard can leave, Birdman bites down hard on the knuckle of his right index finger. The kinkajou curls its legs and tail around Jahangard's arm as if it were a tree branch.

Jahangard has no intention of letting Birdman clamp on to him for five minutes the way he did to VanHollebeke. Jahangard grabs the kinkajou's small head and tries to pry him off. Birdman bites down harder, sinking his canines deeper into Jahangard's knuckle and palm. Get the hose, he tells the Georgian. She grabs the hose but when she turns it on, it runs dry. There's a kink. She runs back to unravel it. Meanwhile rivulets of Jahangard's blood stream down his arm and drip on the cage floor as the kinkajou bears down. The hose gushes. The Georgian aims it at Birdman's eyes and mouth, blasting the honey bear. Birdman does not budge.

They give up on the hose. Jahangard squeezes the back of Birdman's head with all his might. Finally, the kinkajou's jaws ease. Jahangard throws Birdman down hard to stun him, then turns for the door. Birdman stays put on the ground. As he exits the cage, Jahangard worries that maybe he's hurt Birdman. He makes a beeline for Nutrition, dripping blood through Primate Gardens as he goes. He scrubs the puncture wounds with Betadine. Then he hurries through the zoo, out the front gate, and across the vast parking lot to the Health Center.

There, a nurse eyes the wound and orders Jahangard to soak it in Betadine for twenty minutes. She writes him a prescription for antibiotics and sends him for an X-ray. His knuckle has been punctured and crushed. Luckily, no stray bone fragments are found knocking about. All Jahangard has to worry about is an infection.

That is a big worry. The bacteria that thrive in the kinkajou's mouth now call Jahangard's knuckle home. A bite from a kinkajou isn't likely to kill you, but an infection very well might.

The nurses at Moorpark College see cases that few, if any, other college health centers do. They treat what have become run-of-the-mill ailments on college campuses—the eating disorders, depressions, and STDs—plus camel bites to the breast and puncture wounds from parrots. They see fungal infections and bug bites that would be more at home in a tropical country. The EATM students keep them on their nursing toes. "I love it. It's so out of the normal that it's fun," says Sharon Manakas, the center's coordinator.

The students are frequent visitors with normal and abnormal problems. As I sat in the lobby waiting to meet with Manakas one afternoon, in the course of about thirty minutes four EATM students came by. One first year stopped in because she thought she'd broken a finger. When she couldn't see a nurse right away, she decided to soldier on with her lame finger.

In recent years, Manakas has gotten the Health Center more involved in EATM. They have a letter ready to go with any EATM student headed to the emergency room, explaining that animal bites should not always be sutured. The nurses give the first years checkups and have them fill out a long questionnaire. Depending on what they find, the nursing staff keeps an eye on some of them. They will pull students in for meetings if they are repeatedly bitten. Manakas rightly worries that there are many wounds she never sees, because the students don't report bites because they don't want unsafe credits. Even small rat bites are prone to bad infections. A nick on Linda Castaneda's hand from a prairie dog got infected and took a month to heal.

It's not just the bites that worry Manakas but the way the program physically and mentally grinds people down. She wishes the students took better care of themselves, but the program teaches them that the animals always come first. Manakas sees herself as a counterbalance to this emphasis. Don't get her wrong—Manakas loves animals. Over her desk, she even has a black-and-white photo of Bob, the zoo's beloved, deceased water buffalo, with a tire hanging from one horn. She just doesn't like animals more than people.

Sometimes this has put her at odds with the EATM staff. After a macaque nearly ripped the eyelid off a male student, Manakas pressed for the monkeys to be removed. The injury was bad enough, but more worrisome was that macaques are possible herpes B carriers. There is no reliable test for herpes B, and the macaques can shed it without any apparent symptoms. The chance of catching herpes B from a monkey is slim, according to a National Institutes of Health report, but if you do, write your will. The injured student had to be repeatedly tested, which put the nursing staff at risk of contracting the virus as well, Manakas says. The surgical team that mended his eye was also in danger of infection. Manakas thought it too great a risk for a public teaching facility. The Wilsons, who had hand-raised the macaques, fought to keep them but lost the battle. The monkeys moved out.

Animal bites are a constant danger at the teaching zoo, but the real threats are the infections that can bloom so easily. Bacteria are far wilier than the wildest animal. In fact, the students needn't be bitten to get an infection. One student contracted a blazing infection from cleaning the turtle pool with a scratch on her hand. The teaching zoo is rife with bacteria, no matter how the students clean and clean. That is why Dr. Peddie harps on hand washing. There's a picture of the vet looking unusually stern over the sink in Nutrition with a caption inked in: Dr. Peddie says, Wash your hands. Most students comply, which makes for endless cases of eczema. Then infections take root in the cracks in their hands.

The day after the bite, Jahangard's hand puffs and reddens. Despite all the Betadine and the antibiotics, his knuckle is infected. This is especially bad news because an infection in a joint can wreak havoc, causing nerve damage or morphing into a bone infection. Jahangard goes to the emergency room, where he is hooked up to an intravenous drip of antibiotics. The next day his knuckle blooms bigger, redder, and angrier. Back he goes to the ER, where they switch the IV drip to a stronger antibiotic, Vancomycin. Jahangard returns a third day for another bag. The swelling stalls. He returns Sunday for one last bag. He's sent home with two different oral antibiotics. There is no nerve damage. Though it will be tender for weeks, Jahangard should have full movement of his index finger.

The student abandons his plan to train Birdman to walk on a leash. Now his goal will be "to not get chewed." The Georgian has decided not to ask for Birdman. That's no surprise, but her logic is. "He had to hurt the kinkajou to get him off of him," she tells me. "If that was me, I'd let him chew my arm off. I'm not at that point where I could hurt an animal to save myself. So I thought I shouldn't ask for him."

Dr. Peddie worries the Health Center did too little for Jahangard. He thinks they should have gone in with bigger antibiotic guns from the start. He invites the nurses up for a tour of the zoo, during which the vet describes the various bites its residents can inflict, the size of their teeth, and the bacteria that live in their mouths. In the Reptile Room, he tells them that any snake bite should be x-rayed. The reptiles' teeth break easily and can be left behind in a wound, where they will fester. Near Schmoo's cage he describes fish-handlers' disease and how the small scrapes and cuts the students get on their hands while rinsing the mackerel can get infected. As they tour the Mews, the vet explains how the eagles can lock their talons, which can cause a crushing injury. In Parrot Garden, he mentions that a macaw can break a broomstick with its beak. The Health Center nurses eat it up. This is why they love EATM.

Dolphin Dreams

Today is a big day for the aspiring dolphin trainers among the second years. They will make a pilgrimage to Mecca, aka SeaWorld. There, childhood dreams of passing the day in a snug wet suit with a smiley cetacean are a reality for a chosen few. Moreover, if you want to be a dolphin trainer, this is where most of the jobs are; with three SeaWorlds and various amusement parks, Anheuser-Busch employs more marine mammal trainers than anyone. That explains why one second year, who's wanted to be a dolphin trainer since she was five, rises early to carefully curl her long blond locks. She wants to look her best, just in case. Who knows, maybe some SeaWorld honcho will discover her.

This weeklong field trip in mid-February started at Have Trunk Will Travel, where Dave Smith showed off his newly acquired elephant training chops, then proceeded to the San Diego Zoo, where the second years each patted a cheetah on the head. This morning, the group slips through a security gate at SeaWorld to meet up with a mustachioed, seasoned trainer. As expected, he goes on and on about how hard the job is. He's preaching to the converted. What the second years want is the inside track, how to get the impossible—a job.

He became a dolphin trainer in 1976, back when people just happened into the profession. Times have changed. Basically, no one can just fall into the job now. He tells them to get scuba- and CPR-certified, to join the International Marine Animal Trainers Association (IMATA), and to get a college degree. In interviews, he tells them, don't say you want to only work with dolphins. You may have to work with sea otters, even walruses, to get your foot in the door. Don't mention any spiritual connections you might have with cetaceans. "If someone says in an interview, 'I was a dolphin in a previous life,' uh, they're out of here," he says. Give yourself any edge you can, because, "everyone wants to be a dolphin trainer. They'll kill you for it."

The second years follow him past what looks like a big water purifying plant, with huge, bulbous tanks that hum loudly, and through a doorway to the edge of a perfectly turquoise pool, as shiny as a jewel. Nanook, a white beluga whale, bobs in the middle. A bright red ball shimmies across the water. The scene looks like an abstract painting, the colors are so strong and pure and the shapes of the pool and the whale so simple and clean. To one side of the pool, two smaller, grayish females patiently rest their rounded chins in the hands of a trainer. Nanook likes to boss them around and lets loose with a sharp shriek like a seagull's at the girls. "We have to ignore that," the trainer says.

Belugas may be more striking than dolphins but they are not as acrobatic. They can neither jump nor spin. The trainer commands Nanook to get on a poolside scale. The whale gives it a halfhearted try, barely getting his rounded chin on the scale before sliding back into the pool. Then he hoists his great alabaster hulk on the scale. He jauntily tosses his tail up as the numbers tick higher and higher. The beluga weighs 2,000 pounds. Thus establishing his grandeur,

Nanook sinks quietly back into the pool, becoming a ghostly white shadow just beneath the broken blue surface.

Every class at EATM has at least a few wannabe dolphin trainers. This may seem odd, given the fact that the teaching zoo does not have a dolphin. In fact, the only marine mammal is the temperamental Schmoo. Gary Wilson has long dreamed of erecting a small marine mammal facility at the school, but EATM can hardly afford Schmoo's freezer full of mackerel and squid. No tanks are in the offing. The lack of marine mammals, though, does not deter students. EATM teaches you the principles of operant conditioning, the common language of dolphin trainers. Lots of alumni have landed jobs at SeaWorld parks, aquarium shows, and the Navy's International Marine Mammal Project in San Diego. When Brenda Woodhouse attended the school, she was not assigned to the zoo's sea lion. Still, she got a job as a dolphin trainer. The same is true for Wilson, who worked for the Navy. Of the ten or so second years who want to train marine mammals, only two are assigned to Schmoo. The rest make do training flipperless animals.

An EATM degree gives them an edge, but it is no guarantee of a wet suit. As the SeaWorld trainer says, competition is brutal. You might have an easier time becoming a Hollywood star. There are far more dolphin trainers than there are jobs. The International Marine Animal Trainers Association has 1,200 members spread over thirty-five countries, and not all of them work with dolphins. The field has grown some, thanks to interaction programs, but the profession remains tiny. Tom La Puzza, spokesman for the U.S. Navy's International Marine Mammal Project, regularly gets calls from parents of children who have decided on dolphin training as their future career. "They'll say, 'I know it's a hard field to get into,' " he says. "They have no idea."

Why is this? Simply put, because humans are gaga over dolphins. New Age types see an enlightened old soul in the dolphin's smiley gape. Parents plunge autistic, deaf, and terminally ill children in with dolphins, as if the animals had healing powers. Tourists pay big bucks at interaction programs from Tahiti to Mexico to frolic with the cetaceans. At SeaWorld, an average of 100 people a day in the summer plunk down $150 each to get in a pool with the park's dolphins. Often there are tears. An entertainment lawyer, a big guy over six feet tall, bawled his eyes out in the dolphin pool recently. "It's emotional because they've always had this fantasy and suddenly they are living it, and they are overwhelmed," says SeaWorld trainer Suzanne Morgan.

Visitors at the park regularly hold their babies out over the dolphin pool, not to mention the killer whale pool. A family once begged Morgan to let their child who was dying of cancer pet a dolphin before he left this world. Morgan caved and arranged for the family to have some private face time with the dolphins. When the family showed, the dying child was a bubbly, rosy-cheeked baby, and it was the adults who seemed keenest to pet a dolphin. It wasn't the first time Morgan had been sold a line.

No wonder we are so infatuated. Most people have been introduced to dolphins via the stage and screen, which have hyped the animals' charms. Americans of a certain age can still hum the theme song from *Flipper* or mimic the daft lines chirped by the cetaceans in *The Day of the Dolphin*. If you visited SeaWorld as a kid, chances are you left with high-flying dreams of swimming with dolphins or riding a killer whale. Erin Ford, an EATM student from Baltimore, says that "when I saw the trainer come shooting out of the water on the whale's rostrum, I knew that had to be me one day." She was eleven.

As animals in captivity go, dolphins are not such a troubling sight. Instead of a motionless zoo animal behind bars, dolphins frolic in sparkling blue water. We look at the zoo animal and think prison. We look at the dolphins and think swimming pool. Even their trained stunts look more like sports than circus tricks. They are exotic, almost otherworldly, yet familiar. Dolphins live in groups, like us, and with that ever-present smile, they look happy and friendly. They have an athletic grace we find irresistible, not to mention big brains that always wow us humans. "People assume that they are smarter than we are," says Al Kordowski, a longtime trainer at SeaWorld in Orlando.

Dolphin trainers acquire a good deal of their mystique by association. The trainers appear to hold godlike powers when, with a simple flick of the wrist, they command this sleek creature to flip higher than an Olympic gymnast. People will look at a tiger trainer and think that guy has balls. They look at a dolphin trainer and think that guy has brains. "You put on a wet suit and it's like you are a superhero," says Dave McCain, a trainer at SeaWorld in San Diego.

Like their charges, dolphin trainers have traditionally been fit and good-looking. They resemble young college coeds working part time on summer break. They have the bronze skin and bleached hair of surfers and lifeguards. Though many dolphin trainers are essentially entertainers, they have the aura of science about them, unlike circus or movie trainers. Dolphin trainers have their own jargon, bandying about terms like *delta, conditioned stimuli,* and *variable reinforcement*. They are an elite: "There are lots of doctors, accountants, and lawyers," Morgan says. "We're a small, tiny club."

Dolphin training, however, is not the perfect job. For one thing, dolphins bite. They can also ram with their rostrums hard enough to break ribs. Peeved, they can push a trainer out of the water or pin him to the bottom of the tank. In *Lads Before the Wind,* Karen Pryor writes of dolphins' whacking trainers on the head with their tail flukes, blasting them with sonar "you could feel in your bones," and swimming headlong at trainers, only to stop just shy of ramming them.

Dolphin training is just plain hard work. "It's lugging buckets. It's squirting poop," says Morgan. You have to be a crack swimmer. To get a job at SeaWorld, you must swim 220 feet, half of it underwater, and dive 24 feet to retrieve a small weight. Rain or shine, dolphin trainers are outside. All day long, they jump in and out of 52° saltwater pools. The numbing water makes their ear canals swell. Their knees, shoulders, necks, and backs ache. They are prone to sinus infections. Kordowski says trainers swap tricks on how to blow the water out of their sinus cavities.

Dolphin training is especially unglamorous at the start, painfully so if you are a college graduate. Dolphin trainers begin their careers in the fish room, rinsing and weighing bucket after bucket of slimy mackerel, squid, and the like. They work weekends and holidays. They earn a pittance. Even after having paid their dues, they will never get rich. Kordowski says a marine mammal trainer is lucky to make between $25,000 and $30,000. "You can't be head of household and have this job," Morgan says. The guys who stay in the profession, she says, are largely unmarried or gay. Then why do dolphin trainers persist in such trying, underpaid jobs? The answer is obvious. "There are so many cute faces," Morgan says. "It's so self-perpetuating. It just never gets old. . . . It all comes down to that you can get in a pool with a dolphin."

Dolphin training is a relatively recent phenomenon. The first dolphin show in this country was held in 1938 at Marineland in St. Augustine, Florida. It wasn't until the late fifties and early sixties that dolphin training picked up a head of steam, inspired in part by a growing scientific interest. Though Aristotle observed that dolphins gave live birth and suckled their young, little was known about the animals well into the twentieth century. As Dr. Sam Ridgeway writes in *The Dolphin Doctor,* between 1963 and 1973, during which the Apollo program landed a man on the moon, three species of dolphins were studied for the first time.

At Marineland of the Pacific, the world's biggest oceanarium when it opened in 1954 near Los Angeles, zoologist Ken Norris pioneered the study of echolocation with the park's dolphins. In 1962, the U.S. Navy studied the dolphins' knack for diving to great depths without consequence. At the base on Point Mugu in Ventura County, Dr. Ridgeway recruited Wally Ross, the circus trainer who had worked at Jungleland and taught some at EATM, to train Tuffy, a recalcitrant bottlenose with a crescent-shaped scar down its side. Tuffy became the first trained dolphin to work in the open ocean. He dove over 200 feet ferrying tools to SeaLab.

A couple thousand miles out in the Pacific, Pryor cracked a weighty tome on operant conditioning at Sea Life Park in Hawaii. Her husband had started the park in 1963 as a hybrid, an oceanarium that would support a research facility. He drafted her to train the wild-caught dolphins. She had limited experience, training Welsh ponies and her pet dogs, but found Skinner's principles worked beautifully. In *Lads Before the Wind* she writes, "I could see that, given this handful of facts, this elegant system called 'operant conditioning,' one could train any animal to do anything it was physically capable of doing."

To research captive dolphins, they had to be trained to at least a certain extent. If you wanted to test a dolphin's echolocations, you had to teach the animal to wear eye cups. Training dolphins required an entirely different approach from any other animal. If you did something a dolphin didn't like, it could just swim away. Reins, bits, cattle prods, elephant hooks, and the like were out of the question. There's no dominating an animal you can't get your hands on easily and that can just disappear under water. Dolphins could not be forced to work; they had to be enticed.

Early dolphin trainers went to work with whistles, buckets of fish, and operant conditioning manuals. They came up with other rewards—ice, squirts with a hose, a look in a mirror—so a dolphin full of fish would keep working. Using science, they made training fun for an animal.

Though Skinner's principles were used by other types of animal trainers, even if they didn't call it that, marine mammal trainers were the first to wholly embrace the science. They worked out a vocabulary. They standardized their hand cues. Dolphin trainers began getting degrees in psychology and behavior science. They began using training in a holistic way to teach animals to travel and to accept medical treatments. They explained what they were doing in scientific terms, which still isn't the case with many animal trainers. While movie trainers used *affection training* or circus trainers used *gentling,* dolphin trainers used operant conditioning. Unlike many animal trainers, who tend to be competitive and secretive, marine mammal trainers pulled together as a profession. IMATA was founded in 1972.

All this put marine mammal trainers at the forefront of modern training, where they remain. Three of the most influential trainers, Karen Pryor, Gary Priest of the San Diego Zoo, and Ken Ramirez at the Shedd, are all marine mammal trainers. For the past thirty years, marine mammal trainers have beaten the drum for operant conditioning and, moreover, positive reinforcement. At the Shedd, the trainers don't even have a means to tell their animals no. "Once you have the ability to say no," Ramirez says, "you overuse it." Ramirez and his ilk have demonstrated that you needn't punish or coerce your animals to train them. That has been an eye-opener, especially to dog trainers, who traditionally relied on coercion and dominance. This sunny approach is not only better for the animals, but has made training a lot more palatable to institutions such as zoos and sanctuaries.

Marine mammal trainers, especially dolphin trainers, are often considered prima donnas, especially because of their hard-to-follow lingo. One EATM student told me she thought dolphin trainers were snooty. That's the price you pay for being leaders in your field: resentment. For how much longer will the marine mammal trainers lead the pack? The zoo world is certainly fast on their heels, though they have a ways to go. Certainly bird and dog trainers could give them a strong run for their money any day now. When that day comes—when the marine mammal trainers lose their lead—it will be good news for captive animals.

Throughout the morning, the second years wander the park. They hold their hair back with one hand and poke their free arm elbow deep into the bat ray pool to stroke the strange creatures' smooth, hard backs. They watch as penguins frolic under the chilly spray of a snow gun. They stroke the rubbery rostrums of dolphins. Many of the students have been here before, some repeatedly, and have grown a touch jaded. When I mention that the killer whales and trainers are having a play session, one says, "Been there, done that," and heads into the gift shop. Meanwhile, the second year with the curled hair slips off to a quiet spot to call and confirm the date for her swim test at SeaWorld Miami.

Everyone gathers for lunch, where a number of students order a child's meal, because it comes in a plastic holder shaped like a dolphin. I can't help watching one student, a pretty, but very slim one, feast on only the mountain of french fries from her kiddy lunch. She douses them with ranch dressing. Afterward, everyone plops down on a soft stretch of lawn to watch the SeaWorld bird trainers. They bring out Ruby, a fluorescent red ibis, and instruct her to land on a few heads. Ruby slips and slides in Brunelli's curly locks until she lands on her sunglasses and perches there. A black vulture chases the trainers around like a puppy with a funny, waddling run.

We head over to The Shamu Adventure stadium, where three female trainers ride killer whales like surfboards, balancing on their black noses, arms spread to steady themselves. Standing atop the whales' noses, they explode from the pool like human cannonballs. As woman and whale reach the apex of their jump, both arch and dive neatly back into the choppy turquoise waters. Several whales rise up in unison, flippers akimbo, their white bellies shimmering in the afternoon sun. The spectacle is like something Louis XIV would have thought up to wow his jaded court. The bleachers are not filled with aristocrats in feathers and velvet but with tons of shrieking kids in shorts, begging to be splashed by these orcas. The whales are only too happy to oblige. Waves wash over children as they rush toward the pool. Amid the splashing, the Shamus' acrobatics, and the general controlled mayhem of the scene, a second year next to me—an aspiring tiger trainer—wells up. "It's so emotional," she says.

The day ends at Dolphin Stadium, which has a kind of Spanish-mission-meets-rocky-lagoon look to it, with a lighthouse mixed in. There are four dolphins (Duncan, Dolly, Sidney, and Buster) and two pilot whales (Shadow and Bubbles). Bubbles is forty-five. Her age doesn't show. During the customary splashing, she smashes the surface with her dorsal fin, producing frothy waves that chase parents from the front rows. Duncan soars in a tidy arc over a rope twenty-seven feet above the pool. The dolphins are so much smaller and fleeter than the killer whales, but their performance is more poetic, even in its sillier moments. They ferry trainers around the pool. They leap in pairs. They flip and spin with impossible ease. They are the ultimate synchronized swimmers.

After all the families empty out, the second years stay behind, and the dolphin trainers come out in their dripping wet suits. Small puddles collect at their feet. Though the trainers are experienced public speakers, they stand so the EATM students must look into the sun. Everyone has their hands to their brows to shade their eyes. The real distraction, though, is in the pool. As a senior trainer, a super-composed blonde in her thirties, talks about *bridging* and *secondary reinforcements,* the dolphins sail past in the pool, going round and round like an underwater merry-go-round. One animal, a stripe of white down its side, dreamily swims by upside down, its eyes squeezed closed. A pilot whale cruises by over and over with the same little dolphin tucked by his side. One dolphin finally pokes his head out, leans his chin on the pool's glass wall, and stares at the group. No one can resist returning the look. Who can pay attention to anything when a dolphin is eyeballing you?

February

Salsa, the Catalina macaw, won't fly across the stage. Hudson, the beaver, misses his cue. Curly, one of the three mini potbellied pigs, trots on and offstage a couple of times before his trainer scoops up his little black body. Sadie, the sheep, baas loudly backstage. Only Abbey seems to have her part down, lounging on stage, pink tongue unfurled in a quiet pant.

The two casts of "Welcome to the Jungle," the centerpiece of Spring Spec, have been rehearsing for a month. Every morning, one of the two casts takes over Wildlife Theater, and for a good three hours these aspiring animal trainers channel all their overachiever energy into putting on a show. There's something very summer camp about the whole endeavor, yet a lot rides on the production. Spring Spec raises a goodly chunk of money for EATM over its three weekends. Last year, the event snagged nearly $40,000.

Brenda Woodhouse and Jena Anderson, the director of this cast, sit in the bleachers with notebooks in their laps. Anderson has a degree in fine arts and has directed a church choir, so she knows her way around the stage. She's one of the few in her class who does. Most of the students are new as it is to public speaking and now here they are acting.

This Monday morning's run-through is the first time the animals have joined the second years on stage. There has been confusion enough just among the humans, many of whom still don't know their lines or blocking, but add in the animals, and pandemonium rages. Students forget to get their animals, and most animals don't know their trained behaviors. When Rio, the blue and gold macaw, is supposed to appear, someone says, "Rio's in the show?" While many of the animals won't even come out onstage, the pigs can't get enough of the limelight, especially Hamilton, the not-so-miniature Yucatan pig, who lumbers on at every chance he can get.

"Is the rabbit coming?" someone yells.

As with all things EATM, the script is ambitious, which explains why the second years still struggle with their lines. Surprisingly, the text calls for far less from the animals. They are bit players mostly, making cameo appearances as they run or flap across the stage. In the school's old days, the animals were the stars of Spring Spec's precursor, Family Circus. Back then students dressed up in glittery circus costumes and ran ponies with headdresses in hypnotic circles. When circuses became politically incorrect, the school switched to the more neutral edutainment approach. For this show few are being trained to do anything flashy. Students are teaching two parrots to wing into the bleachers and collect a dollar bill from an audience member. Progress has been slow on that stunt. Student trainers spend most of their time just desensitizing the animals to all the distractions onstage.

That is obviously easier said than done. Up until now, the second years have worked their animals in the relative peace and quiet of the teaching zoo grounds. The animals used in the educational shows typically take the stage one at a time. Here, they must follow commands in a vortex of distractions: actors in costume running by, audience members laughing loudly, and other animals onstage. This explains why during rehearsal so many of the animals freeze, as Mazoe, the serval, does. The big-eared cat turns her eyes slowly around the stage and then to the few of us in the bleachers. Paco, a white-nosed coati, is more at ease, walking neatly on his leash. Then Precious, the boa constrictor, squeezes a second year's arm so hard she forgets her line. "I'm getting constricted here," she says.

As Becki Brunelli stands stage right with Malaika, the nervous Nellie parrot, perched on her raised hand, a piece of scenery behind them tilts dangerously forward. "Watch out!" a universal cry breaks out just as the fake wall tumbles, and Brunelli barely steps out of harm's way. She turns to take the bird offstage, but Woodhouse stops her. "Keep it positive for Malaika," she calls.

Nick, the miniature horse, charges onstage at a full gallop, his hooves kicking up dust, his mane flowing. "Whoa!" Woodhouse yells from the bleachers. Nick composes himself and turns a pretty circle and then bows. Hamilton struts back onstage with his trainer in pursuit. When Salsa refuses to spread her wings a second time, her trainer says, "Salsa, fly, damn it!" Olive ambles out on a leash, a banana peel in one hand, looking a bit uneasy. "Come on, sweetie," her trainer Chris Jenkins coos. The run-through of the hour-long show has taken two hours.

"How's it look, Jena?" someone asks Anderson in the stands.

"We'll see," the director answers.

It's mid-February. Though the calendar technically declares the season winter, there are early signs of spring. Legend's meals have shrunk; she doesn't need as much to eat as temperatures rise. The sun crests the mountains to the east earlier and earlier. The students can leave their headlamps in their lockers during the morning cleaning.

The teaching zoo's various newcomers have settled in. Lulu's sides bulge obviously now. The pregnant camel is docile and lovely and everything her son Kaleb and certainly Sirocco aren't. Nick *looks* as if he were pregnant. Despite the efforts of a new set of student trainers, the horse has gained weight, ballooning to 310 pounds. Walter, the kid water buffalo, has become a teaching zoo favorite. He so loves being brushed that he sinks onto his side with bliss. If you rub your hand over his hide, your fingers smell like goat cheese afterward.

Mohelnitzky continues to tinker with Abbey's diet, adding more and more wet food, heaps of ramen noodles, and cans of chicken broth until the dog's meals are a soupy heap. The student adds Mylanta to her evening meal and crosses her fingers. Mary VanHollebeke is not training Starsky for a grade this semester, but she hasn't given up on touching the cavy. She still spends hours each week convincing the big-eyed critter to come close. It's beginning to pay off. He'll pretty reliably eat from the spoon taped to a stick if VanHollebeke holds it very still as she sits in his cage. The student shortens the handle ever so little, so that without realizing it, Starsky is inching toward VanHollebeke as he nibbles on pellets.

Schmoo is less aggressive these days, but still flashes her black teeth on occasion, keeping everyone on their toes. Cain quit puncturing Brunelli recently, but then started sinking his beak into another second year. Samburu, the caracal, hooked a second year's middle finger. She was teaching him a touch behavior, meaning she touches him and not the other way around, as it turned out.

Jahangard spends most of his day with Rosie, the baboon, leaving little time for Birdman, the pit bull kinkajou. That's fine with Jahangard. "Not that I don't like the kinkajou, but I don't like the kinkajou," he says. Still he has to train him for a grade, pulling Birdman from his den box, which seems like a small order until you look at Jahangard's hand. His knuckle is still puffed and sensitive.

Another group of first years begin training rats, toting the rodents to Zoo 2 so they will get used to the classroom. While Wilson lectures, the rats suck noisily from their water bottles and rustle in beds of shredded newspaper. They squat and pee on the tables. Students quickly dab up the small amber pools, as nearby someone eats an overstuffed pita sandwich or a tub of yogurt. A rat occasionally tumbles to the floor or a fight breaks out. After one scuffle, Dr. Peddie patched a wounded rat together with Super Glue.

Linda Castaneda got two hairless rats, hoping their baldness would not trip her asthma. They have. She's so allergic to them she can't sleep at night. She sits up coughing and wheezing. Because they are hairless, the rats can't sleep outside. They'd get too cold. "I almost got in my car at 2 A.M. the other night so I could sleep," she tells me, her voice decreasing to a whisper by the end of the sentence as she runs out of breath. She's returning the pink critters. After she does that, she's going to the Health Center for some antidepressants, she says. Her urge to do things just so combined with EATM's endless demands have stretched her to cracking. She's hoping some pills will purge her of perfectionism. "I'd like to see what life is like without all the noise in my head."

Most of the first years are keyed up over animal assignments, which will be announced in a couple of days. The first year from Argentina has broken out in hives. While emceeing a show for a class of grade schoolers, another first year suddenly walks offstage and throws up from nerves. An older student wakes in the middle of the night from a vivid dream in which she got the serval. Only Chandra Cohn's mind seems elsewhere. The financially beleaguered mom doesn't fret because she doubts she'll get any of the animals she asked for. She didn't have the grades, especially not for Ebony, the raven, whom she longs for. Besides, she's got bigger worries. She always does.

Cohn, the lean, long-faced first year who bawled loudly over pulling her pigeon, counts her pennies each month. She's paying for her groceries with food stamps. She often forgoes lunch, seeming to survive off soda and cigarettes. This isn't the way EATM was supposed to go. After she graduated from high school, she landed a job with movie trainer Bob Dunn. She got pregnant and that changed her plans. She married, had a second child, and worked as a secretary. Over the years, Cohn applied to EATM five times and got in once. Her husband wouldn't let her go, she says. This time he encouraged her. Then last summer, not long before school started, he told her he wanted to separate.

She rented a room in a house where nine or so other students live. Then the younger of her two daughters moved in with her. Her landlord raised the rent. Most nights—after she's helped her daughter with her homework, done her own, and cared for her ferret, two rats, and two hermit crabs—she gets to bed about 2 A.M., which explains why her eyes are always lined with red. Now her ex has sued her for child support.

The funny thing about Cohn is that you'd expect EATM would push her over the brink. It has the opposite effect on her. While the other first years grouse about the program, the animals get Cohn through the day. Here, she can find some peace of mind from the unpredictableness of life beyond the front gate. "I hear the parrots and primates when I come in and it just makes me happy," she says to me while watching Savuti, the hyena, lounge in his cage.

Cohn's observing him for Wilson's Behavior class. She's set her cell phone to ring every thirty seconds. When it chimes, she notes what the hyena is doing, which looks to me like sleeping. "No," she says, her voice lifting, her face brightening. She's discovered there's a difference between when he's resting and when he's sleeping. "Resting, he occasionally lifts his head or moves; sleeping, he's out cold."

Savuti is one of the zoo's biggest surprises. He's enthusiastic, generous, creative, funny, and—something I never thought I'd say about a spotted hyena—lovable. Not only is he one of the best trained animals on the compound, he is the most eager. He gets overexcited with his trainers, trying to anticipate the next command, offering three behaviors when one will do.

There are a few tricks to training Savuti. You work with him through the cage. Second, you must never forget that he is a hyena. That means always watching where your fingers are. Let's put it this way: Savuti cannot have bowling balls with finger holes, because he can break them apart with his jaw. Next, you have to learn the long list of his commands. If the animals don't keep doing their commands, they forget them. That's why animals at the teaching zoo are forever "losing behaviors," as the staff says.

This semester, April Matott and Carrie Hakanson are assigned to Savutes, as the students often call him. Matott is training him to press his left side against the cage bars. Hakanson, whom I find by Savuti's cage later in the day, is twenty-nine and a self-described tomboy who worked with howler monkeys in Costa Rica. She is teaching the hyena to spin but not for a grade. This is the kind of thing EATM students do just for fun—teach a hyena to pirouette. Another student has taught Kaleb, the camel, to dunk a basketball this semester.

"Maybe not a three-sixty," Hakanson says, looking at Savuti, who stares back, his head cocked, "but something dramatic."

Hakanson has worked with two of his existing commands, jump and circle, saying them in quick succession so he is in constant motion. The hyena stands close to the bars, waiting, his big eyes glued to Hakanson. She warms him up with a jump, a sit, and a fetch. She asks him to hold still while she pinches his shoulder through the cage. This is for an injection behavior. "Walk it!" she says. Savuti throws his front paws up on the bars and walks sideways on his hind legs down the cage. Then the second year says

"Spin!" He turns a bit to the left and looks at her with a question in his eye, as if to ask, "Is that it?" She says, "Spin!" This time he tosses himself up in the air like a clumsy dancer and comes down hard with his chin on a log. Unfazed, the hyena rushes back to the bars for a chicken neck, which Hakanson gingerly proffers through the bars.

Late afternoon at the teaching zoo is magical. This is when students tend to get the animals out for a stroll. Calls of "Pig here!" or "Coyote here!" echo across the compound. The students sprint alongside the crazy emu, Julietta. They plead with Spitz, the serval, who lies down every few steps, to please get up. They unreel Scooter's leash so the capuchin monkey can scale her favorite tree, the mulberry near Clarence's enclosure. Walter and Dunny's trainers pull on their leather gloves and promenade with the bulky twosome. If they get too far apart, Dunny stops and Walter cries. The students have to get staff approval for who is out when, because you don't want, say, Legend, a wolf, and Sequoia, a mule deer, running into each other.

This afternoon, Mara Rodriguez steers Spirit, the cougar, down the back road. Spirit is the larger of the two cougar brothers and his fawn-colored coat darkens around his mouth. Though they are well into the semseter, the students aren't ready to take the leash yet. Three of them shadow Rogriguez and the cat.

Spirit stops to smell where Walter peed. He jumps a bit for no apparent reason. Spirit hasn't been on a walk recently. He seems a hair skittish. "If that cat drags me, jump on my chain," Rodriguez says. The students nod. Spirit did drag her once. Rodriguez had noticed just the slightest change in the cougar's gait. Ten steps later, Spirit bolted. Rodriguez threw herself flat out on the ground, spreading her feet, making herself as hard to drag as possible. Still, Spirit pulled her down the back road, scraping the skin off her arms, ripping her shirt and pants. "I'm lucky," she says. "This is such a dangerous field. What has happened to me isn't major, but the major stuff could kill you."

As the posse rounds Nutrition, they come upon a maintenance guy changing a tire on a Bobcat. This clearly unsettles Spirit, who doesn't like cars. They retreat up the back road. Spirit digs in his paws by Sequoia's cage. Rodriguez stops, and the cougar nuzzles her leg. "Are you having a bad day?" she asks. When she turns to go, he freezes, then moves to go in the other direction. Rodriguez bends down on one knee and pulls the chain in link by link so it is nearly taut. "I'll get him to come to me," she says. Once he does, she stands and walks the way Spirit wanted to go. You always want a cat to think his idea was, in fact, your idea.

Eventually, we turn into the section between Hoofstock and Big Carns, where there is a small outdoor theater, and close the gate behind us. Rodriguez orders the cougar up on a stump, and he sits neatly atop the log, his tail draping down the side. She leads the cougar over to where the ground is cushioned with reddish cedar chips. Here, Spirit stretches out on the rich-smelling mattress of mulch. Jena Anderson steps behind the cat, leans down, and starts massaging his shoulders. "Go Jena," another student says, then points out that Anderson's ties from her hoodie are loose.

"You have to know what you are dangling," Rodriguez says.

Rodgriguez hands the leash to Anderson, who clutches it while the other two students take turns stroking the lounging cat. This is another step along the way to learning to walk the cougar. Next door, Legend howls long and hard as C.J., the coyote, leaves her cage for a walk. A low bark rumbles from Schmoo's direction. The scene is deceptively peaceful, five women gathered around a sleepy cougar in the late afternoon sun. Rodriguez knows better, that the calm could break suddenly. She gazes down at the cougar sprawled near her feet and says, "Sometimes I wonder if I'm the biggest idiot but I just think I have something to share that is so neat."

The day finally arrives. On a drizzly morning, the students file into Zoo 2 for the Wednesday meeting. The first years are like kids on Christmas morning, giggling, sitting on the edge of their seats. Announcements, such as "Do not take your rats to the school cafeteria," are kept to a minimum, then the second years are chased from the room. This is, finally, the first years' moment.

Woodhouse hands a folded sheet of paper to every first year, but tells them not to open them just yet. She ran out of herbivores this year and there were more requests for Wendell, the goat, than ever before, she says. The first years nervously fiddle with the pieces of paper. "I'll just keep talking," she says with a laugh. "Anyway, I hope you appreciate every animal you get."

The fluttering of paper fills the room, followed by a few seconds of quiet as eyes scan the long, dense list and then cheering breaks out. First years jump out of their seats and hug each other, yelling, "I got Puppy!" or "I got Cowboy!" Two first years assigned to Laramie, the one-winged eagle, jump into each other's arms. "I'm going to pee in my pants," another first year calls out. Wischhusen dials her girlfriend on her cell. "I got Taj and Zulu," she says.

The Argentinean who broke out in hives stands stunned: she got the animals she asked for, including Savuti, despite her less than perfect grades. Cohn did not get her favorite, Ebony, the raven, but she did get C.J. Patch, Fidone, and Castaneda all got their requests. Castaneda has been assigned to her beloved cougar brothers and the bovine duo, Walter and Dunny. "I got all my first choices," Castaneda says. "It pays to work hard."

Does it? Before the first years leave the room, there's muttering about a student who's been assigned to Schmoo. She dropped Diversity last semester to keep a 4.0. She is also assigned to the zoo's young gibbon. The consensus is quick and sure. She doesn't deserve them.

Larissa Comb sits amid the cheering and carrying on and sinks. Around her are students with lower grades rejoicing, trilling. The straight-A student, a generally all-around accomplished person, got only one of her first picks, a pig. She did not get Schmoo or Goblin, the baboon, or Rio, the macaw. Not that she doesn't like her animals, but she can't help feeling all her hard work has gone unrewarded. She knows why: that one morning she was thirty seconds late last fall.

Combs isn't the only deflated person in the room. The second years rushed into Zoo 2 to congratulate the first years. Mohelnitzky can't find anyone to hug. No first year has been assigned to Abbey. That means nobody asked to work with the dog.

Outside a light sprinkle falls. The morning's rehearsal is canceled. The primates hate rain. The humidity bothers Baxter's arthritis. The second years wander off into the gray, wet morning. The trilling quiets as most of the first years settle in for Dr. Peddie's Animal Nutrition class. He's in classic form. Today's lecture on the minerals animals need in their diet begins with the vet describing his grandmother's goiter, holding his hands up to show just how big it was. "It weighed four to five pounds," he exults.

"Ugh," someone moans in the back of the class.

Over the next hour and a half, he describes a sphincter as "like a drawstring purse," draws a strange-looking cow, which prompts a student to ask, "Dr. Peddie, is that *T. rex*?" He tells them a ruminant's stomach is about the size of a fifty-five-gallon drum and it's full "of a sludge of dead bodies," like a "thick soup." He recounts how he tried to pull a loose tooth from an elephant with the help of a screwdriver and a flashlight. He couldn't pry it free. If you don't pull the tooth, when it falls out, the elephant swallows it.

Quite some time later, after a long journey through the pachyderm's organlike digestive tract, he explains, it appears in the animal's hefty stool. The elephant trainers, he says, gave him the tooth for Christmas.

"Which end did it come out of?" someone yells. His hearing aid must be turned low, because the vet returns to his *T. rex* cow on the board and scrawls some curlicue innards.

By the time class ends, a list of what they call the extra animals, the animals that no one requested and that still need student trainers assigned to them, has been posted outside of the main office. The first years gather round. There are all the usual suspects: the emu, the alligator, the kinkajou, and Starsky, the cavy. There, second from the top, is Abbey. Seeing her name, a tall student next to me says, "I didn't come all the way from Colorado to train a dog."

What these students do every day is controversial. Same with the staff. That EATM even exists is controversial to some extent. That is now true of any animal park, circus, zoo, or aquarium. Though their dreams of working with animals may have been born in the innocence of childhood, animal trainers now find themselves on the front lines of a fight they didn't pick. No matter how good their intentions, they are the declared enemy of the animal rights movement.

That hasn't always been the case. When Gary Johnson bought his first elephant in 1970, he never imagined that one day protesters would heckle his elephant rides at the Santa Ana Zoo. It never occurred to his wife, Kari, that as part of being an elephant trainer she would testify to a congressional committee.

The animal rights movement has been on the offensive for quite a stretch. In response, zoos, animal parks, and trainers have often offered a muddled, halfhearted defense. If the tables are to turn, animal trainers and keepers need to bone up on the legislative system and hone their PR skills. At the very least, they need to think out their position, to have a well-reasoned rationale for why they do what they do. It's not only the animal rights movement they will have to answer to but also any Joe Public that asks them why a big cat sleeps so much in its cage. You don't have to be a PETA member to feel uncomfortable about animals in captivity. Michlyn Hines says she meets people all the time who hate zoos. Becki Brunelli recently had a first date with a man who told her the teaching zoo's animals should be set loose, that they'd be better off squashed on the LA freeways. "I don't think we'll go out again," she says.

There isn't much time in the day at the teaching zoo for philosophy or philosophizing. That is where Animal Ethics comes in. The class is taught on Wednesday evening from 6 to 9 P.M. during the second semester by Leland Shapiro, a professor of animal science at nearby Pierce College who moonlights at EATM. He's completed two postdoctoral studies in bioethics. When he started teaching the course in the late nineties, he found no other like it. He wrote his own textbook.

In Shapiro's class, the EATM students hear names like René Descartes, John Stuart Mill, Chief Seattle, and Gandhi ("To believe something and not live it is dishonest"). They learn how various religions regard animals—that Judaism forbids castration of animals, and Islam teaches that animals should not be used in entertainment. They learn the difference between Malthusianism and dominionism, between animal rights and animal welfare, and between vegetarian and vegan.

Much of Shapiro's class is devoted to the gray area in a debate that's often been cast as black and white. He tells them that Adolf Hitler was a vegetarian. The Dalai Lama eats meat, per his doctor's orders. He points out that there is a cure for the formerly fatal parvovirus, thanks to medical research done on dogs that were given parvo. Medical research on animals produced vaccines for tetanus, polio, diphtheria, smallpox, and whooping cough. Do the ends justify the means? What moral obligation do we have to animals? To ourselves? These are questions that would make the students squirm in their seats if they weren't so tired.

Tonight he mentions the dilemma of Premarin, an estrogen drug used by millions of postmenopausal women made from pregnant mare urine. The horses are kept in a barn stall for six months a year. Even if they weren't pregnant, Shapiro points out, they'd be kept in the barn through the frigid North Dakota winter. Most Premarin is produced by mares on small family farms, he says. Estrogen can be made synthetically, but it's not quite the same as the estrogen produced by the mares. Are the farmers right or wrong to use the mares so? His point is to get the students thinking, rather than accepting popular ideas without examining them.

The EATM students are of various mind-sets, though it's safe to say that everyone agrees captive animals should be well cared for. Brunelli is a strict vegetarian but would like to be a movie trainer. Another student believes animals should be used only for education, meaning zoos are okay but animal parks like SeaWorld are not. Many of the students object to circuses. Not Linda Castaneda. Anita Wischhusen's worked at a dog pound where she had to put the animals to sleep. Among the many stickers on her car, there is one for PETA.

PETA, the biggest and best known animal rights organization, is a four-letter word at EATM. The animal rights movement, with its misinformation and its clandestine techniques, has put anyone who works with animals on guard, including EATM, and made for a bunker mentality. The school is very wary of anyone expressing animal rights sentiments—so wary, in fact, that when, as an April Fools' prank, a group of EATM students posed as animal rights protesters outside the front gate a few years back, they were all summarily expelled by Wilson. Dr. Peddie says if he suspects a prospective student is an animal rights plant, he removes their name from the lottery. The barbed wire perimeter fence around the teaching zoo is not just to keep animals in but also, as at the Johnsons' ranch, possible intruders out, especially animal rights believers with ideas of opening the cage doors.

Not long into this project I found out just how wary the school is. When a young alumna I interviewed misconstrued my questions, she called EATM. She didn't complain about my line of questions but rather suspected I was an animal rights spy, an accusation some staff members took seriously.

Turns out it's not that easy to prove that you are not an animal rights spy. If you were, wouldn't you cover your tracks? I found myself, a horse racing fan, a zoo supporter, and a devoted carnivore, making a long list of character references, from my book editor to an old boss, who could verify that I am a journalist, not a PETA informer. It was a good lesson that brought home to me just how nervous people in the animal industry are, and understandably so.

Suzanne Morgan, the longtime trainer at SeaWorld, told me of a day at the company's Ohio park when the employees were alerted to be on the watch for animal rights protesters. The company had gotten word that something was afoot. What, exactly, nobody knew. Would they try to release the animals? Would they vandalize the park? Hurt the workers? The employees, easy to spot in their uniforms among the milling visitors, felt like sitting ducks, Morgan says.

The movie trainer Hubert Wells has gotten death threats repeatedly. One caller threatened to kill him over his work in *Out of Africa*. In the movie, Karen Blixen and Denys Finch Hatton, played by Meryl Streep and Robert Redford, shoot two charging lions. Wells dug a hole and filled it with padding. The charging lions dropped out of sight into the hole, making it look as if they had dropped dead. His only crime may have been in making it look too convincing. "I know you killed those lions," the caller would growl.

As part of the animal show at Universal Studios, a marshal eagle flies in front of a wind machine. During one performance that Mark Forbes emceed, as the bird flapped, "a lady stood up and yelled, 'This is disgusting,' and started running up and down the stands screaming. Security escorted her out." Forbes tried to recover the show by cracking, "Sorry, folks, I told Mom to stay home today." Nobody laughed.

On the flip side, some trainers told me, Morgan included, that the animal rights movement, despite its heavy-handed tactics and misinformation, has kept the animal industry on its toes in a good way. Karen Rosa, director of the American Humane Association's Film & TV Unit, says PETA and the like have worked in their favor. Movie companies and trainers have become more willing to work with her watchdog group. The association's stamp of approval on a movie gives producers some ammo against any criticism from animal rights groups.

Animal ethics is a heavy, even discouraging subject, but Shapiro does what he can to keep it light. Tonight's class includes two student skits. For the first, two second year students stoop behind a table at the front of the room and perform a hand puppet show on the pros and cons of chicken farming, complete with a catchy song, the chorus of which, "More meat, more money," they sing in high, chirpy voices. The whole room cracks up. "Humor can relax people, not to change their minds, but to see the other's point of view," Shapiro says.

Tonight's class ends a little early. The students close their notebooks and stuff them in their backpacks. They trudge into the dark, take a last look back at the peaceful zoo behind them, then slip through the front gate to the parking lot. The animals, the source of so much philosophical debate, so much consternation, snooze in their cages.

* * *

By the end of the week, a list of who has been assigned to the extra animals is posted by the front office. When Mohelnitzky sees who's been assigned to Abbey, she's crestfallen. She approves of one of the first years, but the other two worry her. She overheard them complaining about Abbey, and one of the two is the first year everyone expects to flunk out or be kicked out. Mohelnitzky goes home depressed. She even cries. She asks her apartment manager again if she can have a dog. He suggests she write a letter to the building's owner. With pen in hand, Mohelnitzky carefully words her request, hoping against hope to bring a zoo dog home.

March

Even though they are close, human and hyena, only inches apart, there are bars between them. April Matott, the second year who went to Have Trunk Will Travel back in the fall, kneels just outside Savuti's cage. The bright-eyed hyena sidles up to the bars. It's the first day of March, a Monday. Matott is running on a few hours of sleep. Though Dr. Peddie does all he can to discourage the students from working part-time jobs, some just have to, and Matott is one of those. She didn't get home from her job at the emergency vet clinic until 2 or 3 A.M., then arrived at the teaching zoo by 6:30 A.M.

She's showing a first year, the Argentinean prone to hives, how a command works. Matott's taught Savuti to let her scratch his left side. She balls her left hand and holds it against the cage. She orders Savuti to press his nose to her fist while she digs into his oily coat with the fingers on her right hand. Matott doesn't use a clicker because Savuti's grown snappish recently, lunging at the bars for his rewards of chicken necks. Matott thinks the clicker aggravates this aggression, so instead she just says, "Good."

Holding her left fist against the cage bars, Matott calls, "Target!" She raises her voice to be heard over the nearby racket of men repairing a drain. The machine they are using backfires with a loud crack. In a flash, Savuti turns his head and bites down on the fingertips of Matott's right hand. "He's got my hand," Matott says.

Savuti pulls, looks confused. Matott is confused, too. She thought her fingers were safely on her side of the bars. Let go, she thinks. She pulls back. The hyena hangs on. She calmly tells the first year to get the hose. The Argentinean dashes for it, but the hose is missing. The construction crew has it.

The delicate tug-of-war between human and hyena continues. Matott hears the snap of a bone break. There goes a finger, she thinks. Then, magically, Savuti unclenches his jaw. Matott's hand is hers again.

The students leave the bewildered hyena. Matott asks the Argentinean to look at her hand to see if all her fingers are there. The Argentinean has a blood phobia. Even the words *vein* or *heart* have made her woozy in the past. She braces herself and looks at the small hand before her. She counts five fingers. The middle one looks as if it's been smashed in a car door. To her relief, the Argentinean does not faint. To her relief, Matott has all her digits.

At the Health Center, a nurse examines Matott's hand and sends her to the emergency room for an X-ray. Matott's middle finger is broken. There's a good-sized hole in the top of her nail where Savuti's tooth punctured it. The nail is not likely to grow back. Matott's referred to a hand specialist.

At EATM they like to say that anything with a mouth bites. That's usually easy to remember around the big carns with their huge, shiny canines. That's harder to keep in mind with the cuter, cuddlier animals, and so the students seem to be forever getting nibbled by the prairie dogs or the baby skunks, one of which has been nicknamed Boob Biter. The funny thing about Savuti is, as dangerous carnivores go, he's cute, really cute. Dr. Peddie says that's why Savuti chomps a student every few years. The hyena got one student twice, the vet says. They let their guard down. Matott says that wasn't the case for her. She thinks Savuti mistook the backfire for a clicker and her pink finger for a chicken neck.

So the new month starts with a bite, a bad one, but it could have been so much worse. Matott is lucky not only to have all her fingers, but that Savuti bit her right hand. Matott needs her left hand to man Laramie, the one-winged eagle, in the Spring Spec show.

Preparations for Spring Spec have built to a fevered pitch. Though classes continue and job searches have begun, all anyone thinks or talks about at the school is Spring Spec, essentially a party that lasts for three weekends. The zoo is scrubbed, weeded, and raked. A team of students with paint brushes dab a dense jungle scene on the backdrop of Wildlife Theater. Costumes are stitched, flowers planted. Emu eggs are carefully emptied so they can be sold. Buttercup's paws are dunked in pink and black paint, so the badger can trot across pieces of paper and make paintings to sell.

There are a few snafus. Spring Spec T-shirts are printed, but the first years, being first years, got them wrong. The second years nail down their lines and polish their animals' training. Though the first years seem to do most of the heavy lifting for the event, overseeing countless details, crewing for the show, and organizing tours, the second years, as tradition would have it, consider it their Spring Spec. And being EATM students, they think of it competitively. They want to raise more money than the class before them, their second years.

The day of the press preview arrives. When the morning breaks and the cast convenes, they discover a costume is missing. Someone took it home to wash because Hudson, the beaver, peed on the costume. Beaver pee or not, the show must go on. Julietta opens the show, high-stepping around the stage, her bluish head slightly bobbing. A small flock of homing pigeons wing overhead, a gray flurry, to the top of the bleachers where a box awaits them—no problem. Baxter gamely, if slowly, carries tools on stage. Rosie throws kisses to the audience and leaps into Trevor's arms.

There are a few animal gaffes, but that's to be expected. When the sheep canter out, one starts chewing on a vine at the back of the stage and doesn't exit on cue. Mary VanHollebeke reaches out from the wings and grabs her halter. The jungle backdrop, which the students added details to just last night, confuses some of the animals. The rats try to climb the freshly painted branches. Puppy, the turkey vulture, his black wings thrown wide, glides past his stump and lands on a video camera. The trainers forgot to bait the stump.

Whatever missteps there are, Schmoo more than makes up for. The sea lion is on. For the finale she balances a ball on her nose, waves a flipper, and smacks a trainer on the bottom. She's a whirling dervish. Everyone cheers and claps. By the time Schmoo's finished, she's speckled in golden bits of earth and glittering like a star.

The following day, the teaching zoo throws its doors open for the first official day of Spring Spec 2004. An army of families in gym shoes pushing strollers marches through. Though "Welcome to the Jungle" has been the focus of attention, there is plenty else to do and the crowd fans out across the compound. There is face painting and a coconut toss. You can take a behind-the-scenes tour of the zoo, which includes a visit to Nutrition, where you can sample monkey chow or bird senet and stroll up the regularly off-limits back road. Tour takers are treated to insider information, such as why Taj, the tiger, is tucked in her den box today; she's afraid of strollers.

The second years trot out various zoo residents. Legend steps up on a tree stump. A student trainer hits the play button on a boom box, and the opening strains of Michael Jackson's "Man in the Mirror" tinkle. Legend's eyes widen. The wolf raises her nose to the sky and howls along to the pop ballad.

At the far end of the zoo, down a hill from Primate Gardens, there's Creature Feature. This is where a long list of professional trainers show off their chops. Nearby, two young women construct a combo information booth–playpen. Inside, a young orangutan named Pebbles in a disposable diaper gambols, tossing toys about, draping her long arms around her trainer's neck, and occasionally trying to scale the fence. She's henna red and restless as a toddler. Pebbles is a huge distraction to the EATM students, especially the interaction-starved first years, some of whom cluster around cooing at her when they should be off giving a tour.

"She's a princess," one of her trainers tells me. "She knows she's pretty."

The two women, who work for the movie trainer Bob Dunn, have brought Pebbles along to raise money for a retirement home for primates in the entertainment industry. A $5 contribution will buy you entry into Pebbles's pen. I hand over a donation and step through the fence. Pebbles is far less interested in me than I am in her. So the trainers give me an orange drink and that catches Pebbles's eye. After I tip the drink so she can take a swallow, the toddler orangutan lets me scratch her back. Her hair is coarse and her back flat and solid as a board. An EATM student walks up to the booth and demands, "How come you get to do that?"

Two little girls rush the fence. Pebbles slips out from under my fingers and over to the girls. She's clearly more interested in primates her age. The younger girl, maybe two or three, is Pebbles's height and the two look each other in the eye. Their parents quickly hand over $5, get out their camera, and the twosome steps into Pebbles's playpen. The girls are less interested in Pebbles than in her toy box, which they immediately inspect. Pebbles holds up toys for them to see. The girls poke around, reach in, and grab a thing or two, oblivious to the orangutan standing right next to them. "When will you ever get to play with an orangutan again?" the mother pleads. This once-in-a-lifetime moment is lost on the youngsters, whose open minds have yet to draw a clear line between humans and animals.

The EATM students dash back and forth across the zoo, from cages to tours to the stage. They aren't allowed to eat in public, so some scarf down snow cones as they stride behind the front office. When they get a few minutes to actually sit down, they duck into Zoo 2 and devour whole sandwiches with the speed of a big cat.

There are four performances of "Welcome to the Jungle," two by each cast. The show has a circus, a train wreck, a mime, a villain, a scene from *Gilligan's Island,* a central mystery, and a referee. It's Oscar Wilde meets Steve Martin. It's very inventive and the cast carries it off gamely, but the intricate wordplay and dense plot make for a long show by family entertainment standards. The metal bleachers creak and groan with restlessness. Wriggly preschoolers are lugged out mid-show.

By late afternoon, everyone is beat and, understandably, tempers begin to flare. Susan Patch notices that Adam Hyde is backing up Trevor Jahangard with Rosie, the baboon, when he should be cleaning. Up until now, Patch has made a point of keeping an even keel, of not taking things personally, but something about this makes her snap. She doesn't yell but, as she puts it later, suggests "not very politely" that Hyde should help out. Hyde, with Rosie nearby, does as he's been taught. He checks his temper. Jahangard tells Patch, "Not now." They are both thinking of Rosie, who doesn't bat an eye but is sure to have noticed that a female, moreover an outsider, has challenged her man. In a moment of being too human, Patch has just committed a social offense in the baboon's world.

It's not long before Patch realizes her error. A second year who saw the scene confronts her. Patch feels her up-to-now clean record, her straight A's, and her perfect attendance slip through her hands. It hits the model student hard. Uncharacteristically, she feels like crying, so the moment Patch is done for the day, she runs for the parking lot. Then she hears Holly Tumas call her name across the stretch of tar. Patch turns, apologizes, and bursts into tears as the other students file by to their cars. Hyde goes home and calls all his friends to discuss human behavior, what he calls Bitchfest 2004.

A carpet of green has unrolled over the hills around the teaching zoo. Sunny yellow wild mustard crowds the valleys. The reminders of last fall's blaze, the skeletons of charred trees with their branches raised skyward and the wiry black balls of crisped sagebrush, still dot the landscape but they look less ominous, more incongruous. The spring sun is hot enough that Wendell, the goat, and Baxter, the pig, nap in the afternoon rays. The reptiles take turns basking in the sun in an outside cage by Nutrition.

Second years fill in Buttercup's hole. Before they dump in the two wheelbarrows of dirt, one student sticks her head down the badger's lair, but can't see the bottom. While out on a walk, Mara Rodriguez chucks Spirit quickly under the chin to remind him who's in charge. He's been ignoring commands. As she says, "If you have a cat on the end of a chain who doesn't listen to you, you're in trouble."

Fewer and fewer first years retain their flawless records. Patch was given a learning contract, essentially a warning, for snapping at Hyde in front of Rosie. Like Patch, Castaneda has fallen from grace. On a recent morning, she drove through a deep fog up the 101, arriving at her usual 5:30 A.M. Then she fell asleep in her car.

She snoozed as other students pulled into the lot and headed into the teaching zoo. No one stopped to rouse her. Her cell phone rang at 6:01 A.M. A fellow first year was calling her from Zoo 1. They had just read her name off the roll.

Chris Jenkins, who trains Olive, announces at a weekly meeting that he's gotten engaged. Everyone cries and cheers. So far, Dr. Peddie's dire predictions of relationship mayhem have not proved true. No marriages have crashed. Serious relationships have held steady. Romances have even bloomed. Terri Fidone often sparkles with pieces of diamond jewelry her boyfriend gives her. He regularly sends text messages that he loves her. The trick to keeping love alive at EATM, Castaneda's live-in boyfriend tells me, is that you accept the teaching zoo as the be-all and end-all for twenty-two months. "I hear everything three times," he says. "When I get home, over dinner, and before we go to bed." Oh, and a fourth: "When she calls her mother or her sister." He's taken over the grocery shopping and laundry. He's learned how to get monkey pee out of her shirts.

Dr. Peddie has gotten a huge shiny new red truck. He needed it, he says, to trailer his new boat that he's docked in Santa Barbara. However, I suspect that the vet, being a man's man, just wanted a big truck. He asks most everybody who walks within five feet of him, "Have you seen my truck?"

He's happy to talk about his retirement gear, but the actual leaving makes his voice tight. Not that he's changed his mind, however much the first years plead. It's just that he can't help feeling a bit anxious. His elderly mother telephones in a worried voice and asks how he can afford this. He can easily, but her question eats at him. His father worked well into his eighties. By his parents' standards, "he's quitting." Then there's the retirement party. Mara Rodriguez has invited four hundred people to a picnic next month. When that subject comes up, his eyes shift from side to side and he draws his lips in. "I don't like being the center of attention," he says. I can't help taking pity on him and ask him something about his truck. His face loosens. He smiles.

The Monday morning after Spring Spec, when a pale fog swaths the teaching zoo, Dr. Peddie pilots his other truck, the old green one with all his vet equipment in the back, to Primate Gardens. The primates are due for their annual TB tests. It's stressful for the vet, for the primates, and for Gary Wilson, who has to catch some of them. "I think I'm slowing down," Wilson says.

This testing will be slightly easier because Goblin was crate-trained last fall after the fire. Last year the baboon had to be darted, but this time on a student's command she goes in her crate and pulls the door closed behind her. Her four student trainers, includng Brunelli, then make themselves scarce, so the baboon won't associate them with what's to come. They don't want to lose Goblin's trust.

Samantha, the gibbon, lets loose with one of her alarm calls—a long whoop—as Gary and Cindy Wilson enter Goblin's cage. They pick up the crate and tip it, so Dr. Peddie can inject the baboon with a sedative. He warns the Wilsons, "She's strong enough to break out of there." She doesn't. They set the crate down and wait for the drug to take effect.

Benji, the vervet monkey, is another story. He's new to the zoo and this will be the first time Wilson will net him. The arboreal monkey has turquoise balls and big fangs. When a second year trainer opens the door to his round enclosure, to her surprise, the latch drops off in her hand. With his agile black fingers, Benji had worked off a nut.

She and another second year shoo Benji into his bedroom, the small room at the back of his cage, where Gary Wilson awaits him. From outside, all you can see are Wilson's bare legs, to which a student wonders out loud, "Why does he have shorts on?" Then

Benji screams and lunges, and there's a flash of fangs and Wilson's legs jumping. The small group outside of the cage tense, gasp, and lean forward. Somehow, the horrible-sounding scene ends up with Benji in Wilson's net. "You should have worn pants," someone says to Wilson. "I should have worn a cup," he responds.

Dr. Peddie pricks Benji through the net. By now, Goblin is so drowsy her second year trainers, including Brunelli, have lifted her out of her crate and set her on a table where the vet can examine her. The baboon leans against Brunelli as the vet pricks her for a tuberculosis test, then checks her teeth, and listens to her heart beat. These annual tests teach the students about what's involved in the animals' general care and how they can help as trainers, such as getting Goblin to hop inside her crate. But the loving often supersedes the learning. Brunelli and Goblin's three other second-year trainers gather round the baboon and do what they normally can't—stroke her and snuggle her. They remark on how soft her skin is. They run their fingertips along her arms, touch her nose, and massage the soles of her feet. Brunelli wraps her arms around the sagging baboon and smiles for the camera. First years stand to the side, enviously watching the love fest. Goblin's tongue hangs out a bit. The baboon lets out a soft sigh.

With the first weekend of Spring Spec behind them, the second years turn their attention back to turnovers. They have to hand over their animals to the first years before they graduate in early May. That seems like plenty of time but it isn't, because turnovers in the spring are long and protracted. The first years, what with classes and the all-consuming Spring Spec, have gotten behind on the many steps of turnovers. The second years press them to pick up the pace. A flurry of notes are posted by the front office, listing turnover appointments by various cages. Mary VanHollebeke yells at Castaneda and the other first-year trainers on Walter and

Dunny. When a first year steps out of class, a second year chides him for missing a feeding with the macaw, Max. "I was in class," he pleads. "Until you've observed a feeding, you can't talk to him," she says. "Then you can give him a nut and say 'Hi.' "

"The second years are giving us a ration of shit," says Chandra Cohn. She's behind on all her turnovers, the furthest being on Samantha, the gibbon. The second year on Boots, the opossum, though, has been giving her the hardest time. "I don't want to even talk about the opossum," Cohn says.

A first year on Puppy, the turkey vulture, has worked up to step 4 in the bird's turnover, learning how to wrap your upper arms with Ace bandages. That done, she proceeds to step 5, pulls on thick leather gloves, and picks up the vulture. With his serrated beak, Puppy grabs and twists her arms—thus the Ace bandages, despite which the bird still leaves bruises. He pulls back her heavy leather gloves and bites her bare forearm. Likewise, Laramie, the one-winged eagle, jabs at his keepers' eyes with his hooked beak. Consequently, they all wear sunglasses. Still, the eagle leaves his mark on these trainers; their foreheads are dotted with small scabs. Susan Patch, who's assigned to the eagle for summer semester, has learned how to gas rats and trap bunnies. That step in the turnover complete, she can now say "Hi" to Laramie. At home, she strengthens her left arm so that she can hoist the eagle. She clutches an eight-pound weight in her left hand while extending her arm. Her muscles strain and quiver. Laramie weighs eleven pounds.

Anita Wischhusen whiles away hours by Zulu's cage, getting to know his honor. She does not groom with him, so the mandrill hasn't had a chance to pinch her yet, but she's worried about when the time will come. "My reaction when something hurts is to pull back," she says, which will make him squeeze harder.

Though Tony Capovilla and Adam Hyde were both assigned to Rosie, the turnover has focused on Hyde. Rosie is already plenty comfortable with Capovilla. She even barks hello to him. Hyde she doesn't know so well. Hyde wants to be a tiger trainer. His looming height will work to his advantage with big cats but not so much so with Rosie. He could easily intimidate the female baboon, though he never has that effect on his fellow students. The lumbering eighteen-year-old learns to look smaller and less threatening. Jahangard tells Hyde never to look at Rosie directly but always from the side. When he walks with the baboon, Hyde slouches. He keeps his voice low and calm—none of his usual cracks or cutting up. So far, the baboon flinched only once, when Hyde offered her a treat.

No matter the primate, Cindy Wilson counsels the students to make their acquaintance slowly. Primates, except for us, are not ones to make friends quickly. If the students push too hard, the primates might push back, only with bared teeth. "Nobody at this point should be getting bitten," she tells them in class. "If you are seeing aggression, you need to change what you are doing. You should be building a bond now." That's why all the primates have lengthy, complicated turnovers. Olive's takes the longest, sixteen weeks. Even the turnover for Scooter, the outgoing capuchin monkey, can take months in some cases.

Hierarchy is a big deal to Scooter, as it is to all primates. Carrie Hakanson is the most dominant of the monkey's second-year trainers, so much so that she can freak out the capuchin just by asking for behaviors. If Scooter is on a tree branch above Hakanson and the second year looks up at her, the monkey will crouch down, a submissive posture. The way Scooter's turnover works is that the four first years go one at a time, from the least dominant personality to the most. Hakanson picked the order and, to my surprise and hers, Castaneda is the first to go, meaning she's the least assertive of the four. This is the first time in her five-foot-eleven-inch life she's been described as such. Number two is a similarly tall woman who wears size 12 shoes. Number three is a former 911 dispatcher with a low, gravelly voice. Terri Fidone is number four.

Hakanson collects the monkey from her cage, takes her hand, and lifts Scooter onto her shoulder. The group ambles up the front road and settles on the curb near one of Scooter's favorite trees, a mulberry near Clarence's enclosure. Hakanson directs Castaneda to sit close to her. The rest of us sit off to the side, per her instructions. On her long lead, Scooter effortlessly scales the mulberry. Hakanson points out a lizard on the tree to the monkey. Scooter doesn't seem to see it, then suddenly her little hands flash, and the lizard tail is sticking out of her mouth as she crunches. Everyone wails. Hakanson calls "Fix it!" and the capuchin rappels down the tree, pulling her lead in as she goes, working black hand over black hand, unraveling its various tangles as she descends. Then she bounds on number two's and number three's knees. Hakanson calls her off, because she doesn't want Scooter to get too friendly with them yet. When Scooter lands on my knees, Hakanson leaves her there. I'm inconsequential, kind of an interesting piece of scenery in the scheme of things, which I find slightly insulting.

Scooter eyes me briefly, scrunches her brow. I hold still and avert my eyes. Then she grooms herself, including a couple of grabs at her crotch, as she balances delicately on my bent knee. As I watch Scooter out of the corner of my eye, I accept my low status in this troupe. Sitting next to me, numbers two and three in the hierarchy still grapple with their respective spots in the lineup. Neither is happy.

"I thought I was pretty opinionated," grouses number two, thinking she should be higher.

"There's no way I'm more dominant than you," says number three, thinking she should be lower. Hierarchy is a big deal for primates, and that includes us.

* * *

There's not much to Abbey's turnover, other than learning to feed her, which is no simple matter. Mohelnitzky tried dosing the dog with Mylanta at night, but that didn't seem to solve her stomach problems either. Undeterred, Mohelnitzky now pours the milky potion down Abbey's throat in the morning. She has trained Abbey to run over an A-frame and through a tunnel for the preshow of "Welcome to the Jungle," but if her stomach bothers her, the pooch won't perform.

In Nutrition, Mohelnitzky shows the students assigned to Abbey how to make the finicky dog's breakfast. In the walk-in freezer, Mohelnitzky pulls down Abbey's tray from a high shelf to show the first years. There is a container of scrambled eggs, pureed chicken noodle soup, and a sticky bottle of barbecue sauce. As Mohelnitzky ladles a gooey mess of soup, eggs, and ramen noodles dotted with kibble into a bowl, she explains that she feeds Abbey 200 grams in the morning and 450 grams in the evening. "Give her whatever is working at the time, and currently it's ramen noodles," she says. "You don't have to worry about this dog getting fat. She is very pampered and very spoiled, so you guys need to keep it up."

What Mohelnitzky doesn't tell them is that she is probably wasting their time. Her landlord said she can have a dog, that he'll make this one exception. That done, she asked the studio training company that owns Abbey if the dog can live with her. They also said yes. There is one last step. She has to submit the request in writing to the EATM staff. Again, Mohelnitzy picks up her pen for Abbey.

Thursday morning, Brenda Woodhouse tries to read the roll as if nothing has happened. She sends the students out on the compound as if it were any other day. She and Rodriguez have agreed not to say anything just yet, but they don't know how long they can keep their emotions in check.

When the zoo has been scrubbed from back to front, the sheep pellets in Hoofstock raked up, and the parrots fed, the students file back into Zoo 1, as they do every morning, to check in, to report anything amiss, to pause briefly before rushing headlong into another busy day. They sink into their chairs. After they are seated, four second years arrive and walk to the front of the classroom. Their faces are tight, splotchy, as if they have been crying.

The EATM staff forever reminds the students to expect the unexpected. An animal can lunge at you out of the blue, bite you in a flash, and escape through a door that is hardly cracked. The animals are wild, no matter how long they are in cages, no matter how close you bond to them. Nature has designed them to be unpredictable. And so is life, but who thinks of that when each day a camel might try to knock you down or a hyena might take off a finger? When these are your worries, the everyday concerns of the world beyond the front gate so easily recede into the far, inconsequential distance. But the front gate only keeps the animals in. It does not keep life, with all its vagaries and fundamental wildness, out.

A second year has died. The words stop time in the room, as the students quickly weigh them. Then their heaviness settles, their razor sharpness cuts. The students gasp and let out small cries. Their mouths slacken. They press their palms to their cheeks as if to scream. Megan Thomas, the second year who ate french fries with a pool of ranch dressing at SeaWorld, who trained Hudson, the beaver, to stand up in his stream, who always smiled and never complained, who just turned twenty-five on Sunday, died in her sleep last night. Her boyfriend couldn't wake her. Her roommates, the young women telling them this unbelievable news, tried to rouse her and couldn't. They gave her CPR. They called 911. Nothing changed the fact that this beautiful, talented girl, who dreamed of training tigers in Las Vegas, was dead.

Everyone sobs, hugs, and staggers out of the room drunk with grief and shock and the absurdity of it all. Not a bite, not a scratch, not a kick. Like Sleeping Beauty, she slipped away from them unscathed in the comfort of her own bed.

"Did she have a heart condition?" students ask. Nobody knows. What could snatch a young woman's life in the middle of the night? There is no ready answer. Rodriguez can't help wondering about her weight, how thin Megan was. Michlyn Hines was worried enough about Megan and a few other students that she had called the Health Center recently. The director was on vacation; she just returned today.

Classes are canceled. So are rehearsals and any kind of appointment. A place so full of bustling activity grinds to an unnerving halt. No one knows what to do with themselves, except for Anita Wischhusen, who gets busy in the front office with some computer work and, perhaps because she's old enough to be world-weary, takes a philosophical approach. "It's the cycle of life," she says grimly.

Many of the second years leave for the house Megan shared with three other EATM students and her longtime boyfriend, where her parents are already busy planning her funeral. The remaining students move in slow motion. Some sit and rest their heads in their hands. In the front office, Kage, the vivacious Texan, lays her head in Brunelli's lap and the two women sob. Others wander off down the front road. Trevor Jahangard and Adam Hyde go to get Rosie out for a stroll. Mary VanHollebeke walks back to Zulu's cage, and the kingly mandrill leans his shoulder against the bars for her to groom, something he hasn't done in a long time. Brunelli settles down next to Goblin's cage.

The students are taught that if they are upset or mad, they should not work with the animals, especially the primates. They will be too distracted to notice the small, telling details of an animal's behavior. Still, the staff doesn't stop any of the students. Suddenly, the certainty of the animals—that they must be fed, walked, and cared for, no matter what—is a deep comfort amid unfathomable uncertainties. The day's lesson has been their hardest at EATM yet: that life, like a wild animal, can never be tamed.

Life can't pause for long at the teaching zoo. There's talk of canceling Spring Spec, but that's impossible. Tomorrow the crowds will return for the second weekend of the zoo-wide extravaganza. Both casts of "Welcome to the Jungle" need to rehearse the freshly cut scripts. One cast, the one that had Megan in it, has to figure out how to get along without her. The rehearsal starts and stops as hollow-eyed second years stumble around the stage and, here and there, break down in tears. The show is supposed to be funny, and funny just doesn't feel right, right now. Funny also doesn't work with actors crying onstage.

Meanwhile, a mob of prospective students and their parents and boyfriends arrives. Anyone who has applied to the program must attend one of these daylong programs. To some degree, these meetings are meant to haze prospective students, to scare off the ones who think animals are cute or have dreams of playing with Bambi. Dr. Peddie does the best he can. He tells them about the lion attack, he warns them that the program will destroy relationships, that parents should not expect to see their sons and daughters for the duration. If you have medical problems, you ought to think twice, he says.

Though Dr. Peddie gives it his all, he can't hold a candle to Tony Capovilla. He's volunteered to give a small group of prospective students and their families a tour of the teaching zoo. Each time I visit the zoo, I'm somewhat surprised to find Capovilla still here. No one has quit the program this year, but if anyone does, my bet would be on Capovilla. Though he has hung in and worked hard, he always seems on the edge of leaving EATM and returning to his $50-an-hour plumber's life. He'll say as much, never hiding his frustration with the program.

Capovilla has a high, rounded forehead, which is often sunburned, a strong chin, and an expressive face. He speaks with a tight-jawed drawl and is in his early thirties. His mother offered to pay his way to EATM. Capovilla thought going to the program would be kind of a vacation from earning a living. He couldn't have been more wrong.

Capovilla's tour makes for a rather discouraging but honest introduction, a counterbalance to the breathless enthusiasm of some of his classmates. The prospective students with their sunny expressions clearly don't quite know how to take some of the things he says in his growly twang, such as "You're worse than a slave, because you pay to do this work" or "You have an easy life now. Enjoy it. I wish I did." At the Rat Room, he encourages them to lean in and "just take a little smell." He tells them, "You'll get used to seeing frozen pigeon heads." Still, try as he may, not even Capovilla can burst their bubble. Now that they have walked through the front gate and seen their dreams in the flesh, few will be turned away.

On Saturday, after the crowds leave, the pots of face paint are capped, the costumes are stashed, and the spilled snow cones are hosed off the front road, everyone gathers in Wildlife Theater for a memorial service. One by one, the animals Megan cared for and trained are brought out onstage. For the first time ever, Clarence, lured by chunks of watermelon, lumbers from his enclosure all the way to the theater. It takes the tortoise an hour. Brunelli brings out Malaika and demonstrates what Megan taught the parrot—a wolf whistle. VanHollebeke fights back tears and lugs out a rabbit. Another student, looking down at her feet, ferries out Nova, the horned owl. Megan used to hoot at him, she says. Hudson is plopped in the pool. The parade continues as dusk turns to night. The humans in the bleachers quietly watch the animals as the dark washes over them.

Baltimore

On the first Monday of April a cold wind grazes Baltimore's waterfront, winter's frosty good-bye kiss on spring's young cheek. The tentative daffodils shiver. The chill air ruffles the green grass and teases the harbor into frosty waves. Though the sun shines, it's a good day to hole up, and that's what a couple hundred zookeepers and animal trainers do in a hotel along the harbor. In a spacious, windowless conference room, they pull up chairs at long tables covered with white tablecloths and pour themselves tall glasses of chilled water. They uncap pens and flip crisp pages of fresh notebooks, ready for reports from the front lines of a revolution that promises an even sunnier future than the budding bulbs outside.

This is the fourth annual meeting of the Animal Behavioral Management Alliance (ABMA). ABMA focuses solely on training. As such, the organization is a rare thing, what animal people call pan-species. At this conference, parrot trainers are elbow to elbow with aquarists and tiger keepers. ABMA's singular focus on training seems straightforward enough, but the truth is the organization sees training as just a tool to reach its broader goal—changing the captive animal world for the better. If ABMA has its way, zoo animals won't have to be sedated for routine medical procedures. They won't be hosed to get them to move from their cage to their zoo exhibit. Big cats won't pace their days away. Monkeys won't neurotically pull their fur out.

Though ABMA's vision may sound utopian, it does not call for blind faith; it is based on the sound science of operant conditioning. As one member puts it, B. F. Skinner "is the father of our industry." ABMA knows seeing is believing, and over the next five days this conference is designed to offer example after example of how Skinner's principles can improve not only the lives of captive animals but the keepers' as well. This week's schedule includes talks on training a black mamba snake to press its nose to a target pole, a gator to rocket out of a murky pond on command, and a dart frog to jump on a mini-scale to be weighed. This morning's lineup alone will cover training African wild dogs, nutria, and fish crows.

ABMA was started in 2000, primarily by marine mammal trainers who sensed a growing hunger among zookeepers and other animal professionals for their techniques. Since then, membership has nearly doubled to about 350. That ABMA exists is proof of the seismic shift in thinking at zoos. Not long ago, training was taboo. An EATM alum told me that when she graduated in 1990, if you wanted to work at a zoo you had to keep your training prowess a secret. It would be a huge black mark. At worst, trainers were equated with circus folk. At best, they were considered prima donnas.

The tables have begun to turn over the past ten years, as zoos have realized the practical benefits of training. Dorothy Belanger, an EATM grad and part-time staffer, offers a case in point. When Belanger was assigned to the Sumatran tigers in 1996, she leapfrogged over more senior keepers, which caused some grumbling. Why was she promoted out of line? Because, Belanger says, she had trained a diabetic gibbon to voluntarily offer its arm for insulin shots.

To some degree, this year's ABMA conference will preach to the converted. Many of the keepers and trainers here know at least a bit of operant conditioning. This conference, though, is sure to rev them up and send them home like so many ardent missionaries, which is good. There are still old-school zoo directors and senior zookeepers to sway whom one attendee refers to as "dead weight." The older keepers can be entrenched, says Gary Priest. "It's not what they signed up for."

Zookeepers predominate here, but there are some dog trainers, an ethologist from Sweden, and even a gymnastics coach, for some reason. Most are young, as are the foot soldiers of any revolution. Most are women, reflecting the gender imbalance of the field. Most are in sweatshirts or polo shirts in solid colors with zoo emblems emblazoned on them. They make their fashion statements with their accessories: socks with tiger stripes, sterling silver animals dangling from their ears, canvas bags covered with paw prints.

A clutch of EATM students, all second years, take up one long table near the front. They are here with Cindy and Gary Wilson. He is one of ABMA's founding directors. Most of the group, eleven of them, bunk down each night at a student's sister's house south of Baltimore. They sleep on couches and the floor, which explains how tired they all look. They drive an hour here each morning in two SUVs they've rented with pooled funds.

Though it's early April, the second years are already done with classes for the semester. Their last month of school is devoted to special projects and job searches. This conference counts as a special project. Three of the students will give a presentation later this week on B.E. at the teaching zoo. Tomorrow, Sarah Harrison and Priscilla Carbajal will speak on the evacuation during the fire. This would have been an emotional presentation in any case, but now it's more so. Their close friend Megan Thomas was to give the lecture with them. The smiley young woman who was to speak about the fire that nearly destroyed the zoo has died somehow. That they are here without Megan is a constant reminder not just of her death but of death itself, so amid all the conference's chatter and buoyancy, the EATM students wander sadly. They should make connections and hand out resumes to training bigwigs. Instead, they cling to each other, as people with heavy hearts do, moving through the cheery crowd like a family in mourning.

Karen Pryor, author, pioneering dolphin trainer, scientist, is also here, if only fleetingly this morning, to give the keynote address. Pryor, in her early seventies, has auburn hair cut in a no-nonsense pageboy. She's of average height, has long, lean arms and a low, melodic voice. She has a good sense of humor and an especially sunny smile. Pryor is a natural storyteller and has a down-to-earth way of talking about training.

She is a star in this crowd—a woman who was training dolphins well before many people in the room were born, who wrote a definitive and easily digestible book on operant conditioning, and who saw the broad implications of Skinner's findings for captive animals long before anyone else did. Her central point is this—hang in there. "One way to change your world is to outlive your critics," she says. If you find your zoo administration pooh-poohs training just persist, she advises. Pryor has, and look at her. She's gone from outcast to luminary. "Open hostility," she tells them, "is one of the best signs that you are making progress."

Pryor comes from a long line of ministers on one side and a secular radical or two on the other. She has three children and seven grandchildren. She's been married and divorced twice. Her first husband, who founded Sea Life Park, fired her from her dolphin training post, essentially for insubordination. She was married a second time, to Jon Lindbergh, son of Charles Lindbergh. She lived with him in the mountains outside Seattle for fifteen years, where she caught up on her scientific writing. Now she lives by herself in the Boston area with her two dogs, a border terrier named Twitchett and a German harlequin poodle named Misha. They are clicker trained.

Pryor considers herself a scientist foremost and has been a link between the animal training world and the behaviorists, some of whom have been appalled at how Skinner's ideas have been usurped to teach dolphins to flip. Pryor is well versed in the abstract notions of her field but isn't above teaching a damselfish to swim through a hoop or a hermit crab to yank on a string for its dinner. She is also an optimist, the way anyone who has sparked two revolutions in her lifetime would be. Before she hung a dolphin whistle around her neck, Pryor changed the way people thought about breast feeding with her book *Nursing Your Baby.* Then Pryor fell into dolphin training and became a pioneer in that field.

Pryor hung up her dolphin whistle in 1974 and wrote *Lads Before the Wind* about her adventures teaching the cetaceans. When she moved to New York City with her teeange daughter to be a freelance writer, Pryor assumed her life as an animal trainer was over. Not long after her migration east, however, she was enlisted by Smithsonian's National Zoo to solve some behavior problems, such as why the polar bears banged on their cage doors all day. Pryor taught the keepers what she had learned with dolphins. In turn, she learned that operant conditioning knew no bounds specieswise. Like other early dolphin trainers, Pryor had assumed Skinner's principles truly worked only with marine mammals. She'd just proven herself wrong.

For Pryor, it was a eureka experience plain and simple. Not so for the zoo world. They criticized Pryor for teaching animals "nonnaturalistic" behaviors or harrumphed that she made them into robots. "It's not fun to be ahead of your time," she says. "Everyone in the field thinks you're nuts."

When a tennis coach asked Pryor at a dinner party to recommend a book on how to use operant conditioning on the court, a light went off in the author's head. She wrote *Don't Shoot the Dog!* to explain operant conditioning in the most conversational tone possible, in hopes people from all walks of life would use Skinnerian principles. Pryor wrote the book primarily, she says, "to get people to stop yelling at their kids." The 1984 book did not reach the broad public Pryor had hoped for, but, to her surprise, dog trainers embraced it enthusiastically. The book inspired the clicker-training movement that has taken dog obedience classes by storm. Meanwhile, her book slowly became the basic primer for trainers of all ilks.

Pryor could rest on her laurels but she doesn't. "Is there a point where you can stop pushing?" she asks. "I don't know." She's curious to see what's ahead and just how far operant conditioning can go. In her talk this morning, she mentions in passing that gymnastics coaches have begun to use a form of clicker training with their athletes. They call this training "tagging"; the clickers, "taggers." It is somewhat lost on this crowd, but Pryor has said something rather startling. The dolphin trainer has now set her sights on another species—humans—to see how operant conditioning can better life for them.

The afternoon includes a talk by trainers at SeaWorld in San Diego on how they recruit young visitors to smear peanut butter on the polar bears' boomer ball or let them shoot wads of cream cheese into their enclosure. It makes the kids and the bears happy. Trainers from the San Diego Zoo and Wild Animal Park show a video of their cheetah streaking after a lure. They learned to do so from none other than Cathryn Hilker at the Cincinnati Zoo. A trainer at a dolphin interaction program in Hawaii describes how guests merrily clean the dolphins' long row of teeth with a Waterpik, feel for their dorsal vein, and place eyecups on them. After dinner, the long day ends with a workshop on aggression led by two trainers who started their careers with killer whales, Gary Priest and Thad Lacinak, who as corporate curator of training for SeaWorld's various parks oversees more than three hundred trainers.

The two men go at the subject alternately like stand-up comics or preachers. Priest teases Lacinak about his pet Chihuahua, Peanut. "I think you need your estrogen levels checked," he needles. "I'm not the one who was a hairdresser," Lacinak shoots back, referring to Priest's previous career. Then they show frightening footage of an elephant tossing a keeper across an enclosure. They recount hellfire-and-brimstone stories of trainers getting hurt or, worse, killed. Their message is simple: do not punish your animals. Aggression provokes aggression, setting off a troubling cycle. Every time your animal bites or lunges at you, it makes it more likely it'll do it again. "It's fun for animals to aggress," Lacinak says. "When you win the fight, it feels good."

This is the marine mammal trainers' bailiwick. In the early days of the field, trainers quickly realized aversive techniques just plain didn't work. A lot of early killer whale trainers were hurt, Lacinak says, until they learned never, ever to correct the orcas. Marine mammal trainers figure if they can use only positive reinforcement, every trainer ought to be able to do the same. Not only is it a more moral way to work, it's more effective and safer, they say, and so Lacinak and Priest hammer away. Don't withhold food, they counsel. A hungry animal gets aggressive fast. Don't hose animals. They only become desensitized to it and begin to think of you as annoying or threatening. Don't use a shield with sea lions. "All it does is piss them off," Lacinak says. "Every time you use an aversive you are hurting your relationship with that animal."

This is an ongoing debate in the training world, because there are shades of punishment and what constitutes an aversive technique. Using a bit and reins on a horse is aversive. So is a whip. While many trainers would consider whipping a horse cruel, not to mention pointless, few would dispense with a bit and reins. What exactly is negative can be very subjective. It's safe to say most people here prefer positive training techniques, or they wouldn't bother with operant conditioning to begin with. Some trainers, though, think using only positive reinforcement is impractical if not impossible, especially when working with big, dangerous animals, to which hierarchy matters. How do you convince a pushy camel you're the boss when he digs in his two-toed feet or bites you in the ribs? At the teaching zoo, if Kaleb, the camel, won't lie down, the students pull down hard on his lead or pop him on the nose with a fist to remind him they are in charge. Lacinak and Priest would not approve, and therein lies the basic philosophical rub.

"Get your animals to like you," Priest says. "Don't punish them." Priest knows he has some convincing to do. Keepers and trainers are humans, and humans are primates, and primates are naturally inclined to dominate and to punish when a little encouragement would do. "Changing the behavior of the animals isn't as easy as changing the behavior of the keepers," he tells the audience. Priest should know.

Like Pryor, Priest is a star in this circle. He's a handsome man with a thick, graying beard. He wears Hawaiian-style shirts with loud prints. Though he's deeply religious, he's got an off-color sense of humor. He's come a long way from his early cowboy days of killer whale training. Like so many of those early trainers, Priest ended up in a tank with an orca by chance. While going to college, Priest worked as a grocery checker, a job he loathed. A woman who worked in the human resources department at SeaWorld came through his aisle. Priest, who is extremely personable, struck up a conversation. Luckily, she had a big order, and by the time he'd rung up her groceries, she told him the park was hiring. He'd never been to SeaWorld, but a job there sounded like more fun than ringing a cash register.

He worked the 4 A.M. shift in the fish house. It was 1970. Killer whales had only been in captivity for five years. Trainers were figuring it out as they went. One morning, not long into his employ, Priest's boss called him over. The killer whale trainer was walking funny, gingerly. He told Priest that he'd be riding the killer whale today. "I was so stunned I never had to worry about stage fright, because I was so afraid of being peeled out of my wet suit like a banana and eaten," he says. "My big break was because of his hemorrhoids."

Priest quit school and became a full-time killer whale trainer. He loved the work, but another SeaWorld boss frustrated him, so he left the park in 1975. Priest went to barber school and cut hair for about five years. He returned to animal training in 1983, when the world-renowned San Diego Zoo hired him as a one-person training department.

In his early days at the zoo some keepers, Priest says, "got their noses bent out of shape" over taking suggestions from a killer whale trainer. Then along came Loon, the diabetic drill. Back then, the only way to get a blood test from a monkey was to sedate him. There was no way Loon could survive having that done repeatedly each day to monitor his glucose level. Normally, the zoo would have had two choices: change his diet and hope against hope or euthanize him.

Marine mammal trainers had long taught husbandry procedures like blood draws. Priest thought, why not try it with a drill? He pulled on tufts of fur on Loon's arm while grooming him. He got the drill to stick his arm through a plastic tube and keep it there. Eventually, Loon would offer his arm through the cage for thrice-daily blood tests. This was revolutionary in the zoo world, which hadn't really considered the medical benefits of training. Priest became a celebrity and, more important, a catalyst for change.

The trainer didn't get to enjoy the limelight for long. In 1991, one of the elephant trainers was killed at the zoo's sister organization, the San Diego Wild Animal Park. Priest was recruited to change how the keepers worked with the elephants to make it safer. People are forever saying the U.S. Department of Labor considers the job of elephant keeper one of the most dangerous in the country, more dangerous than mining or police work. That depends somewhat on how you spin the statistics, and sensationalizes something that doesn't really need sensationalizing. The point is that working with elephants is very, very risky. As Priest puts it, "Elephants don't leave bruises. They leave stains."

For well over four millennia, elephant trainers have taken that chance. Traditionally, the animals are worked in "free contact," meaning human and elephant side by side, no barrier. This is how Gary and Kari Johnson at Have Trunk Will Travel work their female elephants. This system is based on dominance and use of the elephant hook. The Johnsons use the hook as a guide and as an extension of the arm, in the way that many big cat trainers use a whip. However, some old-school trainers assert themselves with heavy use of the elephant hook, especially in Asia. To Priest "it's a system based on negatives."

This age-old system also relies on a long, steady bond between human and animal as well as a trainer with years of experience. As the circus trainers who once took up the steadier life of zoo jobs have disappeared and the older generation of elephant keepers retire, the pool of experienced trainers has shrunk. The younger zookeepers taking their jobs don't have nearly as much experience. Zoos can't always afford the three to five years to train these novice keepers, so many go into the elephant yard without being well prepared. When a keeper is hurt or, worse, killed, it is usually a young one, Priest says.

At the Wild Animal Park, Priest advocated that the elephants be worked in "protected contact," meaning through cage bars, and solely with positive reinforcement. Priest says the elephants learned quickly; the keepers, not so fast. At first they ignored him. Then they argued with him. They vandalized his car. In operant-conditioning talk, they gave Priest lots of aversive stimuli—bucketfuls. "It's mighty reinforcing to go in with elephants and be treated like an elephant," Priest says. "It's so cool. I never appreciated the bond between keepers and elephants. Now, they are no longer part of the herd."

Priest persisted and his system prevailed. Ultimately, all the zoo's elephant trainers who had worked in free contact either quit or were transferred, Priest says. The elephants could make the change, but the humans couldn't. More than ten years later, more and more zoos have switched to protected contact, a trend that is still controversial among elephant keepers and trainers. To this day, Priest says, he remains a pariah among free-contact elephant trainers. Regardless, his career has flourished. Now, as curator of applied behavior, he oversees fifty people at the San Diego Zoo and Wild Animal Park. He consults at zoos around the country and the world.

His work is now largely with people, and he misses the "aha experience" that comes with animal training. What's more important, he says, is that he pass the training baton to the next generation. The revolution is young, and there are still battles to be fought, but Priest compares everything to his wrangle with the elephant keepers. Nothing much seems like that big a fight, after that.

The next morning, a Tuesday, the weather shifts dramatically. The chill wind abates and warmer temperatures soften the air. Anyone who threw on a winter coat to walk to the National Aquarium finds herself shedding it. The day starts with a lean breakfast of limp pastries and so-so coffee, but everything is uphill from there. There's the brand new dolphin show in a sunny theater. It's classic edutainment, with a loud, rocking sound track and two large video screens that flash close-ups of the dolphins. Conference attendees wander the aquarium's dark halls, watching spectral rays glide, hammerhead sharks cruise, and puffer fish bulge. In one darkened tank, a giant Pacific octopus curls a tentacle around a jar. When they reach the fifth floor, conference-goers squint into the light of an early spring day. Their cheeks suddenly feel dewy. They've stepped into the aquarium's man-made rain forest, complete with the loud, raspy bird calls of fidgety blue-gray tanagers and golden tamarins romping in the tree canopy overhead. The monkeys, an aquarium trainer proudly tells me, pee on command.

Back at the hotel, everyone settles in for another afternoon of startling presentations. Priest introduces two trainers from The Living Seas, a 6.2-million-gallon aquarium at Walt Disney World's Epcot, by saying they've used operant conditioning on "an animal no one even dreamed of training just a few years ago." He's referring to the aquarium's two spotted eagle rays, Dottie and Lance, who each have a five-foot wingspan. The two rays had become nuisances. Dottie and Lance accosted divers who got in the tank to feed the fish, pinched them with their hard pallets and pulled their regulators out of their mouths. The two rays even ganged up on guest divers in the tank.

This couldn't go on. The aquarists had nothing to lose. They tried something unheard of—training rays. It was slow going, eighty sessions over four months. By the end, the rays were trained to swim to a PVC elbow filled with clams. A diver would hold the PVC feeder away from his body. No more pinching, no more grabbing their regulators. Now, if a ray swims toward an empty-handed diver, he just points to the diver with the feeder. The ray will turn its wide wings and sail in that direction.

In following talks, there are mentions of sharks trained to go to a station to be fed, a parrot taught to stick his head in a nebulizer, and elephants who will stand still for artificial insemination. Then it's time for the two EATM students to talk about the fire. Gary Wilson introduces them. While he stands at the podium, behind him Harrison blinks back tears, and Carbajal fans her face with her hand. Their fellow students sit up straight and smile to buoy their friends. The charged emotions are out of place in what has been up until now a day of wonders, and the audience seems confused. Wilson explains that one of the students who was to give this talk has just died, and a hushed understanding falls over the room.

Harrison and Carbajal gather themselves and in steady voices recount the few days last fall when the zoo stood in peril. Maybe they are distracted from their grief by the memory of another trauma, the fire, only five months ago. They flash pictures of moon-high flames. They describe the initial evacuation, the long night spent ready to decamp a second time, and the changes that have been made since. A fire emergency plan has been written. Animals have been crate-trained. Their composure does not crack until the very end when a photo of Megan Thomas, smiling her huge, bright smile, flashes before the room.

To the attendees, this is a beautiful young girl snatched too soon. To the EATM students, this is their dearest friend, a peer who inexplicably died, the bad ending of what had been their fairy tale at the teaching zoo. They have learned that the power of operant conditioning is not limitless. No training technique can lessen their pain.

Spring

The morning of Dr. Peddie's retirement party, the vet sits down at his computer in the couple's downstairs office to write a speech. He has a bug and doesn't feel well, but what worries him more are all the people who will gather today because of him. Mara Rodriguez told him three hundred have RSVPed for the picnic. He's shocked. It's embarrassing, being the center of attention, and unnerving that his career at EATM, his career in general, is truly ending. He's retiring. He has the new boat to prove it. Who will he be? he asks himself again, the same question that has echoed in his mind for weeks. He chases away those thoughts and concentrates on the black screen before him. Unshaved, in jeans and a sweatshirt, he sits here, a guy's guy, and looking like it just now, and thinks about the women who have shaped his life, who helped him get from here to there. They've made all the difference.

"Do you know what time it is?" one of these women, his wife, Linda, calls down the stairs to his office.

Over at the Underwood Family Farm in Moorpark, Rodriguez

and a handful of EATM students, the ones who volunteer for everything, such as Becki Brunelli, make final preparations for the outdoor fete. They dump sacks of pretzels into bowls. They arrange centerpieces of goldfish bowls and shells on small squares of netting. They toss up a table near the farm entrance and pen name tags. They erect poster boards plastered with photos of Dr. Peddie atop a tricycle, reaching up a cow's rectum at vet school, and reeling out yet another fishing line. Blue tablecloths fly from their hands. The wind has risen. The clouds gather and darken overhead.

As they work, there's talk, as always, of the teaching zoo and of their animals. A first year explains her black eye. A small South American monkey the size of a squirrel pounced on the back of her head. Then it "bitch-slapped me." Linda Castaneda recounts how Dunny, the Scottish cow, recently pinned her against a fence near Schmoo's enclosure. The next time Castaneda walked him, she was so scared her legs shook. Dunny's best friend, the once adorable Walter, the water buffalo, horned her onstage with his ever-growing rack. Castaneda asked for the cow and water buffalo pair to build her confidence for a camel later, but she's thinking she got the order in reverse.

The party is a family affair. There's a petting zoo and hayrides. The farm, on a flat stretch of fertile earth between some mounded hills, is owned by Craig Underwood, the vet's best and oldest friend. It's a pivotal place for Dr. Peddie. This is the farm where the vet spent a bucolic summer during college, when he made a lifelong friend and fell in love with the arid, open landscape that would become his home. He can stand here today and see how right the promise of that summer was. That is, if he finishes his speech and hightails it over there.

About 3 P.M., cars pull into the parking lot. The cars disgorge vet techs, movie animal trainers, and zookeepers. Former students have flown in from all over the country. Students plaster name tags on guests as they file in. Everyone tips back a large plastic cup of beer and catches up. The sky breaks, and the sun bobs out behind blue-lined clouds. In short order, the parking lot fills.

Dr. Peddie arrives seeming a little keyed up, apologizing for being late, though he's not. Linda smiles and looks stunning in a red leather coat. Before the vet grows more nervous, he's swept into a crowd of old friends and acquaintances. "Oh, my god" becomes his constant refrain as he whirls away. As I pass the vet later, I hear him say to a woman, "Sounds like persistent sinusitis to me."

The gathering is a bit of a who's who in the training world. Cheryl Shawver, owner of Animal Actors of Hollywood, is here all in black. So is Hubert Wells. The dapper Hungarian, still handsome in his seventies, has been friends with the vet since the last years of Jungleland. Dr. Peddie calls "Hubie" his idol because the trainer has had such beautiful girlfriends, including one now who is half his age. Wells could pass as a movie star and speaks in declarative sentences like "A Jack Russell is the only kind of dog to have." He has a scar from every big cat except a leopard: "They've all put a hole in my butt." Wells's approach to training is matter-of-fact, almost to the point of being profound. "You have to explain to the animal what you want," he says. "If something doesn't work, try to think of something else."

The younger generation of movie trainers, such as Mark Forbes, toss back beers here, too, and they offer a study in contrasts, demonstrating that a renegade job has become more of a career. They are Dr. Peddie's former students. They have EATM degrees, unlike the self-taught older generation. They also don't have the big personalities. Forbes, who has worked on a long list of movies from *Homeward Bound II* to *Hidalgo*, is more likely to demystify the profession, saying things like "It's not rocket science." With his thick reddish hair and lean frame, he comes across as the guy next door. "In the past, [studio trainers] were larger-than-life characters," Forbes says. "It does something for them to be this character. It never worked for me. It never gets a better shot or gets animals to work better." Many of the older trainers trained big cats and bears—dangerous animals that naturally made them look macho. Forbes is happy with dogs. "[Big] cat trainers say, 'How can you go on the set [with a dog]? People don't respect you.' "

About the time the second hayride rolls out, the sky dims to gray again and the wind picks up a notch. As people settle in with plates of barbecue, napkins take flight like so many white birds. The board of Dr. Peddie photos blows over. The mercury sinks steadily. You can tell who got here when by what kind of coat they wear. Early comers shiver in cotton sweaters and Windbreakers, while the latecomers look comfortable in down jackets. Everyone gobbles, while clinging to a paper plate of chicken and glancing overhead. Sure enough, an icy, hard rain pours.

Everyone crams in with the caterers under a small pavilion. Rodriguez, who spent months planning this party, climbs atop a table and says, "This really isn't what I had in mind." Speakers are pulled up from the crowd onto the table. Lynn Doria cracks about how much hair the vet has lost. Shawver tells the crowd about the time when an elephant got a rock stuck in its trunk and the vet told her, "Let's wait and see if it's there tomorrow."

A fellow vet tells how Dr. Peddie confided in her that he comes up with his speeches while sitting on the john. Craig Underwood drawls, "I think Jim has been retiring for about twenty years." He tells them how Dr. Peddie exaggerates, even about how many pork chops he ate, and how the vet's penchant for the earthy and grotesque changed his family's table conversation so many years ago. "I'm grateful for the summer of 1963," he says, and the two men tear up. "Now I think he should run for political office," to which everyone cheers.

Then it's the vet's turn. Holding his freshly written speech atop the table, he tells the moist, tight crowd about the women in his life, starting with his ninety-one-year-old mother, who told him this morning on the phone "to stand up straight and talk up" when he gave his speech. He thanks Craig Underwood's mother for hosting him that fateful summer. He thanks Shawver for believing in him. He thanks the female vet techs who always reminded him that the animals were flesh and blood, not anatomical specimens. Even the young females at EATM get a nod for hanging on his every word. "It was like I died and went to heaven," he says. Last, he comes to his "queen," Linda, who kept him from becoming "a cantankerous old vet, treating old dairy cows," and the only important woman in his life, he says, "that I've slept with."

Then the rain stops, and the crowd fans out a bit. A rainbow stripes the sky. Dr. Peddie exhales and says, "I've had nightmares about today, but I've thoroughly enjoyed myself."

Up at the zoo, a new bench with a lion's head at one end and its rump at the other has been placed along the front road overlooking the aviary. Megan Thomas's parents paid for it in memory of their daughter. There is no official word on what caused Megan's death, only rumors of diet pills and the like. Students are left to sit on the bench and ponder the mystery.

One parrot bit off another's toe, a sickly prairie dog had to be euthanized, and Todd, the fox, moved in next to Starsky, the cavy. This relocation has unforeseen consequences. The extroverted fox brings the wallflower cavy out of his shell in a way Mary VanHollebeke has been unable to. His new neighbor emboldens Starsky. Now, when VanHollebeke steps into Starksy's cage, she finds the creature less jittery, even confident.

Over the past few months, the second year has methodically backed up the spoon until it is nearly in her hand. One spring afternoon, as the fox frolics next door, VanHollebeke slips the spoon into her palm and freezes. The cavy leans over and nibbles a pellet or two. She can feel his warm breath on her hand. This isn't officially touching him, but it's damn close. Adrenaline pumps into her veins, but she remains stock-still. One celebratory yelp and her months of painstaking work will be undone. Nearly a year since she first began standing outside the cavy's cage, VanHollebeke watches silently as the teaching zoo's most skittish animal eats from her hand.

With many turnovers finished or almost done, there are no more complaints about a lack of animal interaction among the first years. The cross-species contact soothes like a salve. I find Chandra Cohn, who usually is bug-eyed and straight-mouthed with exhaustion, now wearing a beatific smile on her face. "I just groomed with Samantha for the first time," she tells me one April morning. That the gibbon in Primate Gardens ran her fingers over Cohn's thin arms, inspecting her scabs and moles, means the primate has accepted the harried human into her troop. Under the monkey's searching fingertips, Cohn finds her worries recede.

Animal interaction does not do the trick for Terri Fidone. Oddly enough, Fidone, who has weathered the program better than most, finally begins to sag. Though she's kept to herself, spending her free time with her boyfriend, not joining any of the cliques or listening to the gossip, the competition finally wears her down. "Everybody is so concerned about what everybody else is doing. . . . I feel the vibes. People want to know my business," the former cocktail waitress says. When her aunt died recently out of the blue, she was too upset to go for a cougar walk. Another first year remarked on her absence, which irked her. She misses her family in Las Vegas, even her parents' two teacup poodles. For the first time, she thinks of quitting. She doesn't.

Of her four turnovers, Fidone is furthest along with Kaleb, the dromedary camel. Kaleb is personable and will step close to anyone near his corral, as he lazily blinks his heavily lashed eyes. Like all camels', his breath stinks. Students often feed him whole lemons off the trees near the front office to improve his bad breath. Kaleb and his mother, Lulu, are the biggest animals at the teaching zoo. Kaleb weighs more than one thousand pounds. They are also some of the smartest residents, more so than their stolid countenance may let on. Cindy Wilson says camels are like primates, only they can kill you much faster.

Kaleb is well behaved generally, but also sneaky. He plans his "naughty moves," as a second year puts it, where he'll push a student toward a tree or wall. The student trainers change the route of his walks so he can't think ahead. The boy also bloops. This is an EATM word for the camels' kicking their legs wildly and thrashing about. The camels do it when they are scared or agitated.

Over the past year, a number of the student trainers have been hurt while Kaleb blooped. He hit one in the jaw, knocked the wind out of another, and split open the head of a third. Students must learn to anticipate a bloop by learning the signs: he raises his tail, puts his ears back, raises his head, and tenses his body. He also tends to bloop right before meal time, in Primate Gardens and along the mulch path near Clarence's enclosure. Whether it's the bumpy texture of the mulch under his padded feet or the strange sight of the humongous turtle, Kaleb nearly always goes bananas there.

This is where the second years show Fidone what a bloop looks like and how to handle the melee. As Kaleb begins to shuffle in a lively manner over the mulch, one second year pulls down hard on his lead. That cinches the stud chain over Kaleb's nose and lowers his head so she can better control him. The second years also bring Kaleb here to bloop just for the heck of it, to work it out of his system. They call it his happy dance. However, some trainers would consider this rehearsed aggression, meaning that the more Kaleb gets to bloop, the more he will bloop. It's fun for the camel. The thing is, in a weird way, it's also fun for the students. It's inordinately satisfying to overcome your fears and get a one-thousand-pound-plus animal to calm down. That's why Fidone asked for him.

Lulu, on the other hand, is as ladylike as camels come. Susan

Patch has not had to learn about bloops. Lulu hasn't done that once since arriving at the teaching zoo this past winter. Rather, Patch and another first year, Crystal Oswald, a tall, striking redhead with an oval face and long stride, prepare for motherhood. Lulu is obviously pregnant now. She's lopsided even. Her sides bulge so much that a student walking by her says, "I get cramps just looking at her."

The camel weighs nearly 1,500 pounds, 70 of which she gained in three weeks. Patch and Oswald think Lulu could be a momma any day now. They handle Lulu's teats, trying to desensitize the camel to being touched there. Lulu is fidgety, uncomfortable, like any expectant mother. When Patch and Oswald ready the camel for an education show one April morning, Lulu refuses to be brushed, shifting her wide sides away. "She's being a bitch," Patch says, walking after the camel with a brush. Not even white wedges of apple persuade her to hold still for a grooming. The camel takes them in her mouth only to let the chunks fall to the ground.

Later that same day in Anatomy and Physiology class, Dr. Peddie says that when hoofstock give birth, the labor is violent and fast. The young are born feet first and ready to run with their moms. As the female nears delivery, her vulva streams clear mucus and she stops eating. "Lulu won't take treats," Oswald calls from the back of the classroom. Dr. Peddie pulls a lavender bandana out of his pocket and blows his nose hard on it.

"Tell your vasectomy story," someone in the front row pipes up.

"Oh god," he says, shifting from foot to foot in a kind of "If I must" jig and then, maybe sensing that his days of having a rapt audience of young women for his off-color stories are numbered, launches into the tale. And so the EATM class of 2005 becomes the very last to learn that Dr. Peddie's Novocain block didn't work, that, wearing nothing but socks and lying spread-eagle in the stirrups, he realized he knew the attending nurse, and that the doctor accidentally shocked the vet, which made his legs shoot straight out. "I kicked his glasses off," Dr. Peddie says, smiling.

The second years do not graduate until the first Saturday of May, a few weeks off, but there's little for them to do at the teaching zoo in these final weeks. Their classes are done. Many have already turned over their animals to the first years. Mostly, they roam the grounds with their pals, dash off for bags of steamy fast food, and get animals out.

Still, until graduation, they are technically in charge. They remind the first years in so many ways, small and otherwise. They didn't let the first years make comments on a video tribute to the vet that played at his picnic. They have not invited the first years to the Dr. Peddie roast next month. They grouse about how inept the first years are and make dire predictions. "People will be going to the hospital," a second year ominously tells me.

There is some cause for concern. A first year got out the bobcat without staff permission. Another let Dunny, the Scottish cow, out by accident. Are the first years that bad? I ask the staff. No, they say. This is how it goes each year, they tell me. The second years believe only they can run the teaching zoo and that mayhem will break out as soon as they leave. The first years cannot shine, Rodriguez says, until the second years leave.

For now, the first years, like downtrodden younger siblings, are alternately resentful and intimidated. They are also more harried than ever. They effectively run the zoo, though the staff has yet to give them cage keys. They have to ask the second years to open the doors for them, which is a pain. The first years still have classes. They study the structure of the eye and watch slides of ungulates. They listen to Wilson lecture on dominance. He says some people think that hierarchy is maintained by subordinate animals, and not vice versa. As a graduate student, Wilson studied white-crowned sparrows. The brighter the sparrow's crown, the higher it is in the flock hierarchy. He painted one sparrow's crown brighter. As he guessed, the other birds began to treat the one with Wilson's highlights as boss. This lesson is very apropos just now. The first years still kowtow to the second years, though they are essentially painted sparrows. The paint will soon wear off.

That the first years' time is coming becomes obvious at a weekly meeting. The first years drag themselves into Zoo 2 and settle heavily in their chairs. Hollow-eyed, they wearily rest their heads in their hands. Sitting on tables at the back of the room, the second years pass glazed doughnuts, laugh as if at a reunion, and swing their legs. Woodhouse announces that she has cage keys for the first years. They sit up in their seats and cheer. The second years in back quit jauntily swinging their legs. They slouch and look deflated. The zoo will go on without them.

A week after Dr. Peddie's party, not long before the teaching zoo closes for the day, a Saturday, Crystal Oswald strides down the back road with her red ponytail swinging and her hands full of dishes. She looks over into Hoofstock. Lulu lies on her side. Water jets out of the camel's behind. The baby is coming. Now. Oswald calls up to the front office. "Lulu's water broke," she says into the phone. Give her a bucket of water, the staff member replies. "No, the other water," Oswald says.

In short order, a slender white camel dives front hooves first from Lulu's womb and lands in a kind of tumbled, leggy heap on the bare ground. He's white and fuzzy like a lamb. His fur is darker around his eyes and at the very slight curve that is his immature hump. He doesn't stand up to feed, though Lulu pokes his behind with her nose. This worries everyone. Lulu's last baby did the same thing.

Patch, who was touring the Reagan Library with her parents

when the baby was born, arrives and prepares to spend the night at the teaching zoo with Oswald. The baby does not have an official name, but everyone starts calling him Charlie, naming him after a retired engineer who haunts the teaching zoo each weekend. Charlie hiccups and shivers, signs of possible electrolyte imbalance or an infection.

Dr. Peddie is out of town, so another vet checks the leggy baby twice during the night. He says to bottle-feed Charlie. Someone is sent to buy goat's milk at Ralph's. A calf's nipple is found for the bottle. Every two hours they lift him up and hold the bottle to his young lips. The last time Lulu had a baby, she crushed Wilson, so the camel is tied to a corral bar while they feed Charlie. This maddens the new momma. Lulu moans and groans and swings her wide hips around.

By 10 A.M. the next morning, Sunday, Charlie stands up, though he is as wobbly as a drunk. Before long, he makes a sucking face, as if he might know what he's doing. Students and staff try holding the clumsy baby camel up to Lulu's udder, but because she's tied up, she won't hold still. Patch and Oswald suggest leaving her untied, but staff won't risk it, not after what happened to Wilson last time.

By Monday morning, the temperatures rise uncharacteristically for April, zooming toward 100°. If you lean against Lulu's corral, the bars lightly singe your bare hands or arms. Quite a crowd gathers in Hoofstock. Students and staff hover along the corral's edge, eyes trained on a strange, awkward dance before them. Michlyn Hines, looking pale and resolute, holds the baby camel on his feet with Fidone's help. Oswald clutches Lulu's lead, trying to still the restless momma camel. Hines, with one arm hooked under Charlie's small belly, reaches out with the other and grasps a nipple that dangles from Lulu's swollen udder, pulling it toward the baby's face, as if to say "here." This is risky work, but Hines worries that Charlie has yet to get a long drink of Lulu's colostrum. Despite the students' efforts, the camel still doesn't like having her udder touched. Who can blame her, really? Lulu moans deeply, like the sound of a ship creaking at sea, and pulls away. The baby blinks his black eyes. He totters in Hines's and Fidone's arms, as they steer him unsteadily toward his mother.

This quiet, so-far-unproductive waltz seems to go on indefinitely, then a statuesque blonde steps into the corral. She's an EATM grad, a trainer from the studio company that owns Lulu. She takes Lulu's halter from Oswald, who looks miffed. In short order, the calm breaks and there's yelling and dust flying. The blonde pops Lulu repeatedly on the nose and legs with the lead as the camel pulls back and digs her padded feet in. When Lulu resists, the blonde commands her to walk in circles. Lulu groans. The trainer wants Lulu to hold still so she can be milked. Every time a vet who has stepped into the corral gets close, Lulu jerks away. Whenever Lulu does this, the trainer pops her and they tussle. There's more dust and yelling. It's quite a scene, especially at EATM, where the animals are rarely handled this roughly. It also looks worse than it is, if only because so much dust whirls around.

"Oh, the children," someone says. A school group is coming down the front road.

Eventually, the blond trainer wins the day, and Lulu settles down, does as she's told, and holds still for the vet to pull her teats, but he doesn't manage to get much milk out. "You can't hurt her," the trainer says to Oswald, standing by the corral. "You can't let her get away with anything." Oswald, lips tight, doesn't say anything.

Rodriguez steps into the corral to give Charlie a bottle of goat's milk. The crowd disbands, and mother and child are left in peace briefly. Later, when I ask Wilson what he thought of the trainer's handling of Lulu, he says it doesn't bother him. He would always rather use positive training methods, but they take longer and time was of the essence. Lulu had to be milked.

In the afternoon, after a second year suggests that Lulu's lead be held loosely so she can touch Charlie as he nurses (the same idea Patch and Oswald mentioned Saturday night), the staff gives it a try. That done, Lulu, her nose nuzzling her boy on the rump, stands still for Charlie to nurse. Though Charlie now suckles on his own, there are two more night watches. Patch and Oswald stay for parts of both.

Motherhood takes its toll—not so much on Lulu as on Hines, Patch, and Oswald. Hines becomes so exhausted she can hardly string a sentence together. When she proposes to Dr. Peddie that they insert a stomach tube in little Charlie, the vet asks her when was the last time she ate or slept? He buys her lunch and convinces her the baby is fine: "He needs to bond with his mother."

So far, tending a newborn camel has turned out to be far more work than Patch or Oswald expected. They hadn't counted on staying up four nights in a row right before finals. When Patch sits down on Wednesday morning to take Dr. Peddie's Animal Nutrition test, she falls dead asleep atop her essay booklet. Oswald finds she can't spell anything or write in a straight line. When Patch rouses herself, she reads over her test and realizes she's penned some crazy answers as she fell into her doze. She'd written that you defrost fish by sitting it in a warm bath for five minutes, when she knows that you thaw it overnight in the fridge. The baby camel nearly cost her a grade.

Meanwhile, down in Hoofstock, a baby camel latches on to his mother's nipple whenever she nuzzles his behind with her soft nose. The pulse of the zoo quickens, and the cycle of life spins on, immune to exams, job searches, exhaustion, and the other makings of human drama.

Graduation

Most of the second years are far too young to remember the sexual revolution. To them it sounds like a quaint, distant historical period. Still, there's much tittering as we stand before an epicenter of that more-shattering time. For their very last field trip, the second years find themselves before a Tudor-style house with sprawling grounds in Beverly Hills—the Playboy Mansion. They've come to see the animals, and not the party kind. Hefner has what so many animal people dream of—his very own zoo.

He bought the mansion in 1971, the year *Playboy*'s monthly circulation peaked at seven million, his clubs numbered twenty-three, and his two-eared emblem radiated sexy good taste worldwide. In the early seventies, Hefner was at the tip-top of his zenith. Like so many fat cats before him with cash to burn and status to flaunt, Hef built a zoo.

As long as there has been privilege and wealth, there have been private zoos. The Chinese emperor Wen Wang created a zoo he called the Garden of Intelligence just before 1000 BC. About the same time, the biblical King Solomon started the tradition in Israel. Assyrian and Babylonian kings followed suit. The Roman upper class kept stags and wild boar on the grounds of their villas for hunting as well as for display. Menageries declined after the collapse of the Roman empire, but collections were maintained by European heavyweights, such as Charlemagne in the eighth century and Henry I of England in the twelfth. From the fourteenth to eighteenth centuries, any European royal or aristocrat worth his title had his own zoo. Some of these early menageries evolved into the West's first public zoos in the nineteenth century. Since then, private zoos have been a waning tradition but, as Hefner's proves, not a dead one.

Unlike many other collections through the ages, Hefner's is low on powerhouse species. There are no big cats and no bears. Rather, Hefner seems to prefer beauty to power, and so there are lots and lots of birds: dazzling rare pheasants, leggy African crowned cranes, and rounded trumpeters. Though Hefner's zoo of some 100 animals is private, it has some of the serious aspirations of a public one. That's where our tour guide, John Heston, comes in. He's run the zoo for nearly thirty years.

Heston is a down-to-earth high school science teacher who describes himself as "a farm kid from Oklahoma." He has a bit of a drawl from the open country of his youth. He seems slightly out of place amid the splendor of the mansion. Like so many people the second years have met on field trips, Heston did not follow any career path, so he can't offer a lot of pointers for their job searches. He did not attend EATM, though he has two keepers on his small staff who did. Heston fell into the job in 1975 while he was working on his bachelor's in biology. At that point, the collection was not very healthy. The first order of business was to identify all the animals. That done, Heston studied up on how best to care for them. Before long, animals in the collection not only began to live longer but began to breed. Heston moved on to things like behavioral enrichment, training, and conservation.

Hefner, he tells us, bought the house because Barbie Benton wanted a tennis court, and it had one. "She was a nice lady," he says. The zoo was added early on, and some of the birds have been here from the get-go. Heston leads us around the corner of the house, where the lawn falls away in an undulating sweep like a gentle waterfall. Peacocks, their long tails dragging on the turf like the trains of evening gowns, cock their shiny blue heads to look at us. We round another corner and find geese, cranes, flamingos, and ibis leisurely poking through the backyard with their beaks. Hef, Heston tells us, believes in minimum confinement, so many of the birds wander the estate all day.

The second years quickly duck into the steamy and infamous grotto off Hefner's bright blue pool, then follow Heston single file across the plush turf. This prompts the geese to flap off. A trumpeter named Truman falls into line with us. For the rest of the walk, there is a chorus of "Watch out for Truman!" and "Don't step on Truman!" as the fearless bird steps on our heels. The group advances along a stone path over a stream thick with tubby koi. A net draped over the flowing water prevents the birds from eating the fish for dinner. Heston tugs back a heavy soundproofing curtain, and macaw shrieks pierce the air.

The path weaves between roomy cages of spider monkeys, yellow squirrel monkeys, and tamarins. The cages are shaded by trees, and the small zoo has a tropical rain forest feel to it. Heston has a waiver so the zoo can keep different species in the same cages. For example, there are bunnies in with some crested cranes. "It makes their life more interesting," Heston says.

He steps in with a pair of squirrel monkeys. The two climb aboard his shoulders and drape their long arms around him. They came from a Dumpster in Newbury Park, he says. "They are a lot like people but they are incredibly honest." He lifts them from his shoulders to go, and they hover by the door as he steps out backward. "It took six months of training to get them to let me leave."

As we near a cage full of exotically colored pheasants, a crested crane steps in our path, blocking the way, and looking surprisingly menacing for something with such spindly legs. "That's Spot," Heston says, and shoos him off. The bird was raised by hand and ate dog food, he explains. "That's why Hefner called him Spot." We duck into a series of stone buildings. Inside it's moist and close. This is where the hornbills Heston breeds live.

Since the late '70s, Heston has bred a long list of endangered hornbills, starting with Jackson's hornbills. Since then he's raised Taricitic hornbills, Oriental pied hornbills, and black Malaysian hornbills, to name just a few. He is currently working on the large bar-pouched wreathed hornbills. The species, he says, "has one foot on a banana peel." If it weren't for a project like this, he probably wouldn't have stuck at the zoo for all these years. This gives the extravagance a purpose and some intellectual import. Otherwise, the zoo would exist only for one man's pleasure.

There are reminders everywhere of that man's pleasures. Heston takes us over to the gaming room, where a few of the students tackle the roomful of video games. Second years pile into a mirror-lined, pillow-strewn den and smile for the camera. The tour ends back at the shimmering pool where coffee and pastries await. "Fruit salad," someone calls, and the young women with plates in hand surge around a voluminous bowl of cantaloupe, honeydew, and berries. As the EATM students merrily fork chunks of melon, a goose pads across the lawn. A crested crane turns his fancy head to the sky. A flamingo balances on one leg.

The unseasonably hot temperatures persist into May, with the mercury cresting at 90° at midday in Moorpark. The early heat singes the green hills. As they wither to brown, the blackened scars of last fall's fire recede into the dun-colored landscape. At the teaching zoo, Aladdin, the chinchilla, has been moved into the air-conditioned cool of the supply closet off Zoo 1. The sheep and Gee Whiz, the llama, have been sheared. The llama vaguely resembles a Q-tip with his tufted tail and shaggy head.

On the first Monday in May, another round of primates, including Scooter, are given TB tests. I find Linda Castaneda seated in the air-conditioned comfort of Zoo 1 with Scooter mewing in a crate before her. "Take a nap!" Castaneda coos in her Scooter voice, which is high and uncharacteristically girly for her. Castaneda fills out feeding logs while waiting for the monkey to rouse from the anesthetic. She has hives on the base of her neck from letting her rat run around her shoulders. She's been at the zoo every day, she tells me, since January 31. Castaneda has switched from Zoloft, which made her crabby, to Wellbutrin, still hoping to tame her perfectionist ways.

The baby camel has been named Sahara officially, though Susan Patch still often calls him Charlie. In Hoofstock, I find Patch and two other students scrubbing the young camel. One works her fingers through the caked poop in his tail. Lulu stands nearby, her eyes trained on her goofy boy. Her coat has fallen out in big, mangy-looking chunks since delivering her baby. In a corral one over, the freshly sheared sheep look up. Dunny and Walter butt heads lightly. Schmoo suns herself in her corner.

Though the students squirt Sahara, he totters over for more of the same. This is good. The plan is to get Sahara used to handling so that, unlike his mother, he won't mind the touch of human hands. Adam Hyde swings open the gate to Hoofstock. He's been recruited to scoop Sahara up so he can be weighed, though he didn't expect to hoist a wet camel. Hyde cups the animal's neck with one arm and his belly with the other and steps on a scale. Together, camel and first year weigh 358½ pounds. Hyde sets the camel down and steps back on the scale. He weighs 252. "I've lost weight," he exclaims. Sahara has gained nearly twenty pounds in the ten days since he was born. The baby camel now weighs 106½.

The new crop of Schmoo girls file into Hoofstock. They are still learning the sea lion's commands—they write them on their arms—and her fickle ways. They've already begun to sing their commands in clear, high voices just like the second years. Today, they saw her wander for the first time, while she was doing an education show this morning with the second years. *Wandering* is teaching zoo talk for when the temperamental sea lion zones out and ignores commands. Nobody knows why she wanders. Maybe the unseasonable heat puts her in a mood. Maybe a small seizure rattles her. Maybe she is just being Schmoo.

Two of the second year Schmoo girls arrive. "Has anyone showed you sneaky-walk?" one asks. The second years open Schmoo's enclosure, and the shiny dowager scoots out. Sneaky-walk is like something out of a Marx Brothers film, basically a goofy, shadow walk. Schmoo follows the trainer, and each time the trainer stops and looks over her shoulder, the sea lion does the same. Schmoo is not the least bit wandery now, and turns her black head just at the right moment. Her timing is right on, despite her bad hearing and cataracts. The new Schmoo girls try to concentrate, to soak up the finer points of sneaky-walk, but they can't help laughing.

For the first years, this is the last week of classes before finals. One is on academic probation and may soon be saying adieu. In Dr. Peddie's A&P, they take notes on white muscle disease, which leaves the tissue looking like "acid poured on a steak," the vet says, ruining yet another food for his students. In Diversity, they are close to the end of a very long list of species, nearing the scavengers and what is listed on the outline as "most abundant large decomposers," seven vultures and one stork. In Behavior, Wilson hits a philosophical note: "Thinking freedom is a good thing is a very anthropomorphic view," he says.

In addition to the rat maze exam this week, there is the knot test. In between classes, first years sit cross-legged on the small lawn in front of Zoo 1 tying and untying, practicing the thirty some knots Wilson taught them. The teacher prowls the zoo looking for students to test. They duck into supply sheds to hide from him. When Wilson catches up with Chandra Cohn, she shows him the two or three she knows. She has bigger tangles on her mind, such as how she will pay the rent this month. Her student loan has gone AWOL.

This is the last week, period, for the second years. Next week they'll be like any visitors to the zoo, restricted to the front road, unable to open a cage, and forbidden, like a first year, from animal interaction. "Graduation is going to be hard," Mohelnitzky says. "I don't want it to end. I don't want to say good-bye." Another second year tells me that graduating is like getting out of prison: "You think you can't wait to get out but then you don't know what to do." Another tells me after twenty-two months in her uniform, "I've forgotten how to put outfits together."

It's a week of last this and last that. Every time I walk past Buttercup's cage, one of her second-year trainers presses the badger's square, flat body to her chest. Second years throw a leg over Kaleb for one last camel ride. They pull up chairs to primate cages for extra grooming sessions.

Rosie, the baboon, knows the drill, that Jahangard will soon leave her. In her way, the baboon says good-bye first; she quits obeying his commands. When he tells her to get in her crate, Rosie doesn't. A dominant male baboon would let Rosie have it for this infraction—push her or bite her. Jahangard takes the middle road. He has Adam Hyde and Tony Capovilla stand behind him and the three men stare at the uppity female baboon together.

Mary VanHollebeke accepts defeat with Starsky, the cavy, though she came mighty close. She'll graduate without touching the cavy as she had dreamed. After eating out of her hand a few times last month, Starsky never came as close to VanHollebeke again. "I thought I'd make more progress," she sighs by his cage. "I'd planned for him to be out by now." Her efforts have not been for naught. Starsky is less scared of humans, especially a certain human with long dark brown hair. Moreover, VanHollebeke leaves EATM with a newfound patience. As the human accustomed the cavy to her looming presence, the small South American rodent desensitized VanHollebeke to the fast ticking of the clock inside her head.

Amy Mohelnitzky is the only one that doesn't have to say good-bye to her favorite animal. The staff approved Mohelnitzky's request. Abbey will go home with her. The pooch will still belong to the studio company, but she'll live with Mohelnitzky and her husband. Earlier this week, some of the second years threw her a shower. They gave her dog toys, a blanket, poop pickup bags, and food bowls on a raised stand so the briard mix doesn't have to strain her neck when she eats. Mohelnitzky still doesn't know for sure when Abbey can move out of the zoo. Abbey has to stay until the teaching zoo gets another dog.

The second years dread saying good-bye to each other as well. Next week they will scatter. The one who curled her hair for her SeaWorld visit has gotten a dream job at Discovery Cove in Florida. CVP, pigeon puller extraordinaire, will move up to Vallejo to work with sea lions for the summer. Two second years are off to the Point Defiance Zoo in Tacoma, Washington. Another is moving to Portland, Oregon, where she hopes to land a job with Guide Dogs for the Blind. April Matott will drive her ancient Plymouth back to her hometown of Albany, New York, and look for work there.

Not everyone will depart for farther points. Mohelnitzky will teach puppy training classes at night and work as a receptionist for a vet during the day. Chris Jenkins, Olive's trainer, landed a job with an education outreach program in Pasadena. Two other students found work at Animal Actors of Hollywood in Thousand Oaks. VanHollebeke will work at the teaching zoo's summer camp. And Becki Brunelli, to my surprise, has taken a part-time job with an area studio company. Brunelli kept hoping that some kind of job would open at EATM or even at Moorpark College, but one never did. As graduation neared, she saw the job with the movie trainer and thought, why not? Dr. Peddie gave her a rave review.

The thing is, film work probably pays less, at least starting out, than most any other animal work. New movie trainers typically work as cleaners, as Brunelli will. She's being paid $700 a month to clean and care for 35 dogs, 20 cats, a couple of raccoons, and sundry other animals. If Brunelli goes on a set, her hourly wage jumps from $8 an hour to $32, but for as long as she's on the shoot only. The trick is to get on a set. New movie trainers often train an animal new to a facility, in hopes it will be needed for a shoot. The more in demand the animal becomes, the more money that trainer makes. That could take some time and is a bit of a gamble. In the short run, Brunelli will supplement her meager wages with computer work. With her student loans and credit card debt, she'll have to.

The final week is riddled with parties and rituals, like a summer camp winding up for the season. This morning, Thursday, is the willing ceremony, a kind of changing of the guard meets baby shower. It marks the official turning over of the animals from the second years to the first years. As usual, there are presents involved. Everyone convenes in Wildlife Theater, the first years on one side, the second years on the other. VanHollebeke and Brunelli call out various areas, starting with Big Carns. Then students stomp down the creaky bleachers to the stage en masse. Squealing and hugging rule as the second years give halters, chunks of salmon, brushes, marshmallows, handmade key chains, and bags of Goldfish to the first years. Then the first years hand over presents to their second-year buddies. For the moment at least, any bad feelings between the classes evaporate as the air fills with the sound of gift paper ripping and animal print socks and the like being unwrapped.

There's another party Thursday night, and then Friday families alight at various nearby airports. The second years tour them through the zoo and demonstrate the magic they have learned with various animals. The day's mood is merry like a holiday's, between the second years' excitement and first years' seeing the end in sight. That is, until the king tumbles.

Not long past noon, Zulu falls off a platform in his enclosure. A student hears him land with a thump and arrives to find the mandrill on the ground having a seizure. Word spreads quickly, and in short order a number of his trainers rush to his side with worried expressions on their faces. The mandrill sits cross-legged on the ground, something no one has seen him do before. He furrows his brow. The blue in his face pales to the color of threadbare denim. He doesn't smile his big, fanged, clown smile. VanHollebeke and Kristina Nelson, the coolheaded Iowan, stroke his hands and look deep into his mournful eyes.

"Was he fed anything different?" Nelson asks two first years standing nearby. No, they say, nothing was different.

Zulu's owner has been called; his medical care is up to her. The sudents have some theories, that this has to do with his migraines, and that he ought to have an MRI. They hold his hands through the cage and comfort the fallen king. Even in his reduced state, Zulu appears to enjoy all the female attention. He scratches himself and even picks at some dried poop on his butt—small, welcome signs of his regular self. His timing is horrible. It was going to be hard enough for his second-year trainers to leave him, but now, like this, how can they?

"I'm staying here," Nelson says, clutching the hand of the ailing mandrill.

The crush of time is slowed by nonstop graduation festivities, which I'm beginning to think is the point. Friday night there's the Dr. Peddie roast up at the zoo. As people filter in, they deposit flowers on Megan's bench, tie balloons to it, and then join an endless line for burgers. The vet looks chipper, though he reports recurring dreams of traveling and not being able to reach his wife on the phone, from which he wakes up in a cold sweat.

The roast, as all things EATM, is a long and ambitious affair. It begins in Zoo 2 with a slide show. There are photos of Dr. Peddie asleep in his underwear and with his arm up various animals' behinds. Back outside, as the light fades, Castaneda presents him with a nautical compass from the first years. There are funny, touching tributes from the second years, too, though one mentions that Dr. Peddie always teased her about her past visits to bondage clubs and another, the zoo's gardener—the one who loves Sequoia, the deer—tells the bleacher full of parents, younger siblings, and grandparents about how the vet warned him that "he'd screw himself to death." Brunelli, the emcee, looks mortified. "All I can say is, I'm sorry," she says, before asking Dr. Peddie to say a few words. Then the vet relates his story about how, as a joke, he told the dean of the college that he accepts only students with a C-cup or bigger to EATM.

As the evening winds to an end and families and students filter out, a small group remains behind to spend the night with Zulu. For their very last twenty-four hours, a few second years get to do night watch. It turns out to be uneventful. The mandrill sleeps the night away on his platform, rising at daybreak. He tugs on his left ear and scratches his lower back. At about 6:45 A.M. he has a cup of warm broth and throws it up. Close to 8 A.M. he pulls his penis, ejaculates, and eats his sperm. Zulu is clearly feeling more himself.

Not long after Zulu improves, the second years with their families in tow return for another tradition, Family Circus. For two and a half hours, the students trot out their various animals. Schmoo, as always, closes the show, entering stage right with a ball balanced on her upraised nose. Not long into her act, though, the sea lion ignores a trainer when she asks for a kiss. Schmoo is wandering. The trainers cut her act short and, without seeming obvious, usher the star offstage.

The afternoon is one long drawn-out good-bye. It's hard not to feel bad for the second years. The Schmoo girls burst into tears after their last training session. One student walks around the teaching zoo crying her eyes out. Brunelli settles into a chair by Goblin's cage and lets the olive baboon gently groom her hand.

As the sun slowly dips and the heat of the day lifts, the second years gather backstage in Wildlife Theater to line up for the graduation ceremony. "We'll have to ask the first years to open the front gate for us," one grumbles in line. They march out. There are floating candles in the shallow moat around the dirt stage. The bleachers are packed with video camera–toting relatives. The mountaintops that burst into flame last fall glow pink in the early dusk. There's applause and inspiring speeches from staff members and classmates. Jenkins tells the crowd that this is a wonderful place but not an easy place to be. Gary Wilson hands out their certificates one by one. Then the students intone the EATM pledge ending, "I accept the challenges which lie ahead of me. I will represent with pride, dignity, and honor the Exotic Animal Training and Management Program of Moorpark College."

Finally, the moment has arrived. The new graduates embrace their families and friends and hug staff members one last time. If they tarry, no one rushes them. Slowly, they trickle out into the wide world, glancing back as the front gate closes behind them.

The Zoo Is Theirs

The first years' tenure at the teaching zoo begins, as you might expect, dramatically. When finals are posted, the first year who is on probation passes, at which her classmates cheer, but two others flunk out. That brings the casualties for the class to four. One student is not a surprise: she's the one everyone expected to get kicked out, the slacker who made Mohelnitzky decide to adopt Abbey. She missed a C by 10 points in Anatomy and Physiology. She's shown the front gate.

The other first year is a surprise. It's Tony Capovilla, one of Rosie's new trainers. A few months ago that might have been a relief to the former plumber, but not now. Though he's grumbled about the program in the past, he has come around and accepted how consuming it is. Capovilla is stunned. He thought he had a B going into his Diversity final but, given that test score, Capovilla missed a C by a fraction of a point. Dr. Peddie, as one of his final duties before retiring, tells Capovilla he's out, that it's the student's decision when he's done, now or at the end of the week. Now, Capovilla says, and shows himself the front gate. He could reenter the program a year from now but he would join a class with six other men, which means he could not count on training Rosie, the baboon—the reason he came to EATM in the first place.

This means that Rosie now has one trainer, Adam Hyde, and Hyde does not have a backup. The only way Rosie can come out for a walk right now is if Cindy Wilson comes along. When she does, Hyde feels self-conscious, he the primate novice, she the expert. Rosie seems to notice his discomfort and challenges Hyde. When he gives her a command, the baboon looks at Wilson. It's the baboon's way of asking, Who is in charge here?

Within a week, the slacker is back. Dr. Peddie, in what he says is an effort to "be an all-around-bitchin-good guy," cut her a deal, what he thought was a secret deal. Within about twelve hours it wasn't. Nothing ever is at EATM. Going over her final, the vet finds an answer that he could give her 10 more points on. The correct information is there, though not in the proper order. This girl pulls on his heart strings. Her personal life is a mess, he says. The vet will pass her, he tells her, but she must read *Ten Stupid Things Women Do to Mess Up Their Lives* by Laura C. Schlessinger and write a book report. He realizes it was a weak moment. "She is not someone to go to the mat for," he tells me on the phone.

Brenda Woodhouse, newly in charge, hears of Dr. Peddie's "secret" deal from Capovilla, who arrives in her office with all his quizzes from Diversity under his arm. "I'm shocked Dr. Peddie thought anything would stay a secret here," she says. Capovilla meets with Wilson, too, and the two of them go over the tests and find a score that was entered incorrectly, a 26 that should have been a 27. That gives Capovilla the sliver of a point he needs to pass Diversity. In like manner, he rejoins the program. The first years are glad to have him back, but some can't help wondering if he, too, was given a deal, if only for Rosie's sake.

That's just the start of what sounds like a rocky first few weeks. Ebony, the raven, bolts, flying through the legs of the student blocking the cage door. Paco, the white-nosed coati, likewise escapes, and is found trundling around Maintenance. While onstage, Friday, a raccoon, scampers into the bleachers. The students catch up with him on the aviary lawn. Zeta, a turaco, flies off and flaps around the zoo for twenty-four hours before she is lured back to her cage. One student gets docked unsafe credits for forcing Julietta, the emu, on the scale along the back road. Then the same student is bitten on the finger by one of the lemurs. Next she lets Zeta out by accident, which she blames on her bandaged finger.

"It's just frustrating if you are killing yourself for this," Castaneda says. "We'll be judged on their slacker asses." Castaneda has given up on medicating her perfectionism; Wellbutrin gave her a "gnarly" rash and nightmares. She's thinking now, though, given how much all the recent screwups bother her, that maybe the drug worked. When she posts a note about Dunny throwing up by Zoo 1, she tears it a bit. She pauses and considers taking the note down and redoing it. She doesn't have the time, so resists the urge. EATM, with its impossible to-do list, may accomplish what the antidepressants didn't—train her to be less than perfect.

Michlyn Hines struggles to make the teaching zoo run like a professional zoo. She threatens to lower grades if students don't catch up with the zoo's paperwork, reminds them to keep the front gate closed, and referees disputes. Most of the first years get along fine, but the few who don't do so dramatically. "There were no personality conflicts in the class before," Hines says. "Now I have two trainers on a bird who won't talk to each other." She's referring to Anita Wischhusen and a young woman assigned to Zeta, the turaco. Two of the Schmoo girls don't get along either. Then not long after he returns, Capovilla gets into a shoving match with an older male student.

Capovilla had gotten Rosie out of her cage and then called up to the office to see if he could walk her. Technically, Capovilla should have asked before getting the baboon out. The older student, a broad-shouldered man with a soft voice, said so to Capovilla. Moreover, he said it in front of Rosie. Capovilla kept his temper but after he put the baboon away, he saw the older student stepping out of Zoo 1 and he let his rage rip. "He threatened my life," Capovilla says.

The kinkajou, now nicknamed Cujo, sinks his teeth into the third hand in just over a year. Kristy Marson—a star in her class, a Schmoo girl, a student seemingly born to train—ends up with the kinkajou attached to her hand like Jahangard and VanHollebeke before her. The staff rules that students will no longer go in with Birdman. They will train him from outside his cage, just as they do the tiger, lion, and hyena.

Then Ronald Reagan has to go and die. All the mourners pull their cars into Moorpark College's lot and from there are ferried over to the Ronald Reagan Presidential Library, where the former president lies in state. This means there's essentially no place for the EATM students to park. For two days a skeletal crew of students tends the teaching zoo. Now that the zoo is theirs, the first years wonder if they will ever get in the groove.

By late June, I find that they have. The first years have new confidence, as the staff predicted they would. No longer intimidated by their second years, their brows unknit, and joints loosen. Smiles are common. There's a new ease in their step. They have taken charge. Holly Tumas says the class is doing well, "more than second years expected, more than first years expected."

Though they've been training their animals for less than a month, the first years are making quick progress. Sahara, the baby camel, lets Susan Patch and Crystal Oswald put their fingers in his soft mouth. A first year has taught Schmoo to keep two feet of stomach tubing down her throat. All the new cougar trainers have taken the chain while on a walk. A first year, one who aspires to get a Ph.D., has taught Cookie, the loquacious gray parrot, to whistle part of *The Andy Griffith Show*'s theme song.

Marissa Williams, a pretty young horsewoman from North Carolina, has taken over Nick's training, which includes his weight problem. She has yet to take some pounds off the maxi-mini horse, but Nick is fitter. Nick still pesters Gee Whiz until the llama spits food at him, which the horse then gobbles up. Williams figures that's a lost cause and just keeps the horse moving. She's organized teams of walkers to get him out for romps twice a week. She runs him in wide circles on the aviary lawn. She hooks him up to the cart and trots him around the zoo.

Zulu has improved. He flashes me his signature fanged smile while he relishes a wad of gum with big, exaggerated, lip-smacking chews. He hasn't had any more seizures. In the dry summer heat, his butt pad has dried out and cracked, getting blood all over his cage, but, nevertheless, he seems his old regal self. He's groomed with all four of his new trainers—a sign he has accepted them, though he often grabs Wischhusen during these sessions. The first time he grabbed her he pinched her until she got a bruise. Then he dug his fingernails into the bruise until he drew blood. "He's cruel," she says. "He's a little shit."

The mandrill and Wischhusen are in a bit of a power struggle, which comes as no surprise. Wischhusen goes her own way. She's made it this far without pulling a pigeon, mostly because she just doesn't want to. Now she's going against everybody's advice with Zulu. When Wischhusen irks the mandrill, she does not offer her arm to Zulu for a punishing squeeze. She has trouble, not surprisingly, being subordinate. It is not in her nature, which is probably why she is so drawn to Taj. Training a tiger is all about being dominant, which Wischhusen has no problem with, given her broad shoulders and football coach's voice. With Zulu it's the reverse, and Wischhusen cannot bring herself to curtsy to the king. She will not submit by letting him vent his rage on her arm. Wischhusen thinks her approach is worth a try. "They said he wouldn't eat cooked yams and he does," she says.

The irony is that Zulu consequently grabs Wischhusen more than any of the four trainers. Often when she sits close to his cage or grooms with him, Zulu snatches her arms with his black hands and clenches with all his mandrill might. Sometimes it scares Wischhusen so badly it makes her legs shake.

There's plenty of news of the recently graduated second years. A couple have moved to Hawaii and another to Mexico in pursuit of dolphin training jobs. One recent graduate has left movie training for educational outreach work. Another had a change of heart before she even started with a studio trainer, did a brief turn teaching at a dog training school, quit that job, and has started her own dog training business.

Becki Brunelli fills me in over a dinner of Mexican food in a roadside restaurant in Moorpark. She orders chicken fajitas without the chicken. We eat on the early side, as the sunshine just barely slants, because Brunelli's due again on a movie set tomorrow morning at 4:30 A.M. She was there today. The domestic cat she was working with got loose and ran around the set, which got Brunelli in hot water with her new boss. She shrugs her shoulders. Ever the optimist, she says it was a learning experience.

Brunelli herself has adjusted fine to life after EATM. She's as bright-eyed and energetic as always. She's single again, she says with a smile. "Right now, I can't see myself with anyone. I'm not lonely. What I want now is adventure." Her days, though, are spent mostly raking up dog poop and scraping dried pigeon droppings off the bottom of their cages. She has splinters in her fingers. "It just sucks. It's like being a first year again," she says. She worries that her coworker is a teenager with a high school diploma. Brunelli, thirty-four, has two college degrees, not to mention graduate work, under her belt.

As we sip frosty margaritas and a motorcycle blares into the parking lot outside, Brunelli says she and her fellow graduates are working in facilities that are not as well kept as the teaching zoo. At the trainer's facility, Brunelli says, water bowls are filled only if they are empty. At EATM the water was changed every day regardless, she says. I detect a touch of what Dr. Peddie calls the EATM curse. Graduates have gotten a reputation for being know-it-alls and telling employers how to keep their facilities or train their animals. Dr. Peddie warns new graduates to keep their lips locked, but that's not always easy for animal people when it comes to animals.

In Wilson's training class, the first years get down to the nitty-gritty of using operant conditioning. It's one thing to understand the principles, another to put them to work. Wilson says students having trained a rat think they know it all. Obviously, they don't.

One of the biggest hurdles for a novice trainer to get over is being human. Despite our big brains, we have trouble looking at the world through another species' eyes. We're so busy thinking, we have trouble paying attention to what we are doing. We fidget, so our body language is a mishmash of signals that could confuse the most intelligent species. We are naturally bossy and expect human traits like obedience and loyalty from the animal kingdom. We take things personally.

Wilson devotes a chunk of a class to anthropomorphism. The English language lends itself to such, but Wilson warns, "The danger is, if we are being really anthropomorphic, it could lead to bad decisions." He cautions the students against saying things like "The animal is glad to see me" or "The animal loves me but hates my co-trainer." Animals are not emotional. Humans are.

"We want you to care about all the animals, even the rats in the breeding colony, but you can't let animals get away with things because you want to be a friend rather than a trainer.

"You have to be consistent," he says. "You can't be a buddy."

That isn't a hard lesson for Castaneda just now with Walter, the once-adorable water buffalo. No one feels much like petting him these days, though he's still technically a baby. He waves those ever-growing horns around. He digs in his hooves on walks. He tried to ram a sheep backstage. It took five people to stop him, including the sizable Hyde. Castaneda and his other student trainers think the problem is the rather large pair of testicles hanging low between Walter's haunches. Castaneda and his other student trainers want the punk water buffalo "chopped," meaning castrated.

Castaneda assumed that of the animals she was assigned to, the cougars were the most dangerous. Now she thinks it's Walter. She's still shook up over a walk last week. Wilson suggested separating Walter and Dunny, the constant companions, during the stroll. Dunny was led ahead and into the hay barn where Walter couldn't see him. The water buffalo panicked and charged after his furry, bovine friend, pulling a frightened Castaneda along with him. If he were a large cat, she could throw herself down. If he were a camel, she could pull his head down. With a water buffalo, all she could do was run alongside as Walter frantically looked for his pal. The experience taught Castaneda that she needed a pair of steel-toed boots.

Ever since Walter took off, his trainers are uneasy about getting the duo out. A minimum of three trainers is needed to walk "the cows," but these days it's hard to find three willing students. Though she's scared as well, Castaneda worries the twosome isn't getting any exercise. Today, she and three students gather at Dunny and Walter's corral because Woodhouse is going to come along on the walk and offer pointers, which makes everyone feel more confident.

The two bovines smell of citronella. Ash-colored stripes have emerged on Walter's chest and legs. Dunny has put on weight. A student to either side of him, the Scottish cow lumbers out, his big woolly coat a beautiful caramel color in the sun. No problem.

Then Castaneda in her leather gloves and big boots leashes Walter up in the corral and turns him toward the gate. He immediately plants his heels and wags his head with those now enormous horns. Castaneda hangs on for her life, flexing her arms, gritting her teeth, but she's really no match.

"I wouldn't put up with that," Woodhouse says. "Put on a second lead."

A first year with blue eyes and chestnut hair to her shoulders attaches a second lead on the other side. Woodhouse shows the students how to turn the water buffalo. One student holds still, the pivot point, and the other walks him around. It's a simple but effective system that gives the two young women some control.

Walter's slightly better behaved coming out of the corral but then heads for a tuft of grass near Lulu's enclosure, and another wrestling match ensues. The two students dig in their heels and pull back on his head. Somehow they get him across Hoofstock and out the gate, though Woodhouse tells them they should always go ahead of the water buffalo, "so he doesn't pin you."

We start up the back road. Two enormous bovines and six women. We are briefly a picture of pastoral calm, that is, until Walter swings his great horned head toward another patch of grass, and another battle royal begins, with Castaneda and the brown-haired first year pulling back and Walter pushing forward. "Do you ever whack him?" Woodhouse asks.

"Gary discourages us from slapping him," Castaneda says. Woodhouse smiles.

We continue up the back road, amble past Zoo 1 and the office, heading toward the front gate. When we arrive near Megan's bench, about the time Clarence cranes his head our way, Walter steps off the blacktop onto the grass, dragging the students with him. "I'd take him out of there because he's just reinforcing himself for being a pain in the ass," Woodhouse says.

The two women tussle with the adolescent buffalo, try pivoting him, but get nowhere. Then in a flash, Woodhouse like a superhero whacks Walter on the neck with the palm of her hand, grabs his lead, and single-handedly pulls him back on the blacktop. It's a sight to behold. We all look at Woodhouse with silent awe. Walter seems just as surprised as the rest of us.

For the rest of the walk Walter just trots along like his old baby self. We take a spin down the front road and then double back. Gee Whiz turns his head to watch us pass. Harrison, the Harris hawk, squawks at us. Dunny stops to take a long voluminous pee. Castaneda blocks Walter's view. "He likes to drink Dunny's pee," she explains. When they stop by the scales, Dunny weighs in at 1,022 pounds, nearly 50 pounds heavier than he was a month ago. At 800 pounds, Walter has gained 80, but he is still growing. Once the two bovine buddies are put back in their corral, they playfully start butting heads. "You're a brat," Castaneda says to Walter.

Woodhouse says the crack she gave the water buffalo was like a flyswatter's to him. It was meant to get his attention, she says, which it did clearly. The trick is to use it judiciously, she tells them. If you don't, it will lose its meaning.

Like Castaneda, Terri Fidone learns an animal needn't have claws or fangs to be dangerous. She, too, assumed the cougar brothers were the most worrisome animals she'd be working with. In fact, it's a domestic animal, Kaleb. "Odds of [the cougars] attacking are very low," she says. "With a camel, every day it could happen."

Kaleb bloops and bloops and bloops. One time by the aviary lawn, Fidone could not get the camel to stop. Though she pulled hard on Kaleb's lead, she could not get the camel's head down. Kaleb kicked her in the back, knocking Fidone forward onto her knees. When she fell, she lost hold of the lead. He could have stomped her. Instead, he quit blooping and stood still by her side. As Fidone walked him back to Hoofstock she realized she couldn't hear. All the adrenaline coursing through her veins had temporarily turned off her hearing. Her nerves were shaken. Worse, her pride was hurt.

Fidone's not sure why Kaleb's become so incorrigible. She wonders if the second years had him bloop too much. He tends to bloop when she has the lead. Could that be because she's training him for a grade? Wilson, inspired by the recent ABMA conference, has instructed the first years to use more positive reinforcement with Kaleb. They give him wedges of sweet potato for walking well and staying calm, though the treats seem to have the opposite effect on him. He's pushy. Could it be this change in approach? Or is the four-year-old reaching sexual maturity? Only Kaleb knows.

Fidone is no chicken. She's skydived. She rides motorcycles. She's served drinks in the world's biggest casino. She's tough and likes a challenge, even one with a big, fatty hump. Still, there are some days she can't bring herself to get the camel out for a walk. "It's so trying with him at times. You think, am I going to live today?"

There's a new resident at the zoo. You can see him if you stand at the back of Primate Gardens and look between the lemurs' and the binturong's cages into Quarantine. Chances are, he'll look right back at you with such eager, glad eyes it can break your heart. The teaching zoo has a new dog.

Kasey came from a pound in Las Vegas with a case of kennel cough. His age is guessed at one and a half. He's a mutt but looks like he has a lot of Australian shepherd in him. He certainly has the breed's bright, attentive eyes and fluffy coat. He's moved into Abbey's old digs next to Samburu, the ornery caracal.

A zoo is a lot for a young dog to get used to. When Kasey first arrived he barked his head off in his new, exotic home. Now the pooch is relatively quiet. The primates aren't bothered by him, nor is Legend. Consequently, Kasey's easier to walk than Abbey was. He eats only dry kibble, which he gobbles right up. Yet Kasey is no star like his predecessor. The pup knows only one command—Sit! He has no manners. He pulls on his leash and jumps on people excitedly. He's pent up with young energy and easily becomes overexcited. Consequently, two of his student trainers lead him to the front of the line for a campus walk one afternoon. That way, Kasey won't be distracted by the assortment of animals trailing behind him.

This campus walk heads out the back gate and along a dirt road that curves down a hill to the playing fields below the zoo. Larissa Comb holds Nick's halter. Three students, their arms held high and crooked, carry macaws on their hands. Rosie ambles along on her leash next to Capovilla and Hyde. Holly Tumas carries a young gibbon in her arms.

As the group turns to walk out on the baseball field, Kasey starts barking, and his trainers stop to calm him. "That dog is crazy," someone holding a macaw says. The students with their various animals pause briefly and then push on. At first Kasey doesn't seem too rattled by the procession passing him, but then Rosie jumps up on some jungle gym bars. That does it. The dog lunges on his lead and woofs his lungs out at the baboon. Both trainers grab hold of his leash and flex their arms. Even after the campus walk has moved off across the lush, grassy field, Kasey still goes bonkers. His student trainers stand by his side and watch the group walk away. "They won't wait for us when he is having trouble," one says. "If it was another animal, they would wait," the other says. Kasey bugs his eyes at the receding baboon.

The group settles on a cushy carpet of lime green grass that sparkles in the afternoon sun. The athletic field is a brilliant oasis amid the browned foothills. It's not too hot and a soft breeze blows. The occasional pop of a bat hitting a ball sounds. A team practices nearby. Tumas sets the gibbon down, and the monkey joyfully stretches out on her belly on the turf. Rosie carefully picks through the blades and, having found one that is to her liking, plucks it and pops it in her mouth. Two young boys on bikes ride past, never seeming to notice the baboon sitting on the lawn. Comb runs with Nick at her side across the field and then back. Susan Patch does a cartwheel.

Now that the teaching zoo is theirs, the summer promises to be long and busy. For a moment, though, they pause and relish the fullness of a beautiful day, from the sun burnishing their faces to the lemon-scented grass as fresh and inviting as a newly made bed. Before too long the group pulls itself to its feet and, holding leashes and reins, drifts across the carpet of green, meanders up the scrubby hill, and vanishes like a dream.

August

As the summer progresses, the second years' warning proves true: this is the most demanding semester yet. There are only half as many people to clean the teaching zoo. There are still classes to study for, presentations to give, and now animals to train. They are even required to spend a few days volunteering at area zoos. Yet it's so satisfying to work with the animals, to lay their hands on them, talk to them, and teach them, that the first years find they can soldier on through the long days under the inland California sun. The rush of time at the teaching zoo, the endless highs and lows, and the days packed with the unusual and wondrous pull them through their weariest moments.

In July alone, Walter, the water buffalo, is castrated, another first year flunks out, and the capuchins break out of their double-baleen cage. One hot morning, Susan Patch notices something skitter through Primate Gardens, a rabbit, she thinks. Then she realizes it's a monkey. Patch sprints up the front road, throws open the door to Zoo 2, and, out of breath, wheezes to the crowd, "Capuchins out!"

The clever troupe had worked off a bolt or two on their cage. The students get a hasty, hands-on lesson in what to do when an animal escapes. One, Flash, is quickly recaptured, but Elvis, the alpha male, laps the campus. The student from Georgia who took care of fawns climbs atop a batting cage, pulls on leather monkey gloves, sits cross-legged, and waits. Elvis runs across her lap a few times, stopping once to lean his flat nose close to hers. As he scampers over her legs one more time, she grabs him.

Elvis bites down hard on her hand. She holds on to the monkey's waist. He is returned to his cage. The first year pulls off her glove and finds to her amazement the bite did not break her skin. Later, when another monkey pees on her hand, it swells and turns red. An invisible wound has somehow become infected.

The students practice what they learn in Wilson's training class on the zoo to good effect. Castaneda teaches Scooter, a capuchin, to climb into a bucket and the now much more mellow Walter to stand on a mark. Adam Hyde teaches Rosie to ride a skate board and to let him brush her teeth. Anita Wischhusen trains Taj to sit on a fake scale that she slides under the tiger's cage but makes little progress with her novel approach to Zulu. The mandrill regularly grabs her, putting Wischhusen in her place. In fact, she's the most subordinate among the four student trainers. Kristy Marson, working outside his cage, teaches Birdman, the kinkajou, to go into a crate on command.

Marissa Williams teaches Nick to step as prettily as the Lipizzaner Stallions and, moreover, manages to thin the horse. At last Nick drops below 300, to 290 pounds. His middle no longer bulges. He no longer huffs and puffs when he pulls the cart.

Patch trains Sahara, the baby camel, to be haltered and led. That done, Sahara goes for his first trailer ride, to Burbank to *The Tonight Show* with Jay Leno. His mom, Lulu, comes along for the ride. During the rehearsal, Harry, as he is nicknamed, nervously pees a bit on the rubber mats under him, but settles down nicely. However when the full band begins to play, the poor baby camel screams and security men come running. Patch holds her hand over his soft mouth but to little avail. Then, after the lantern-jawed comedian dawdles with a caracal, time runs out for Harry.

Chandra Cohn, no longer able to afford her rent, moves home with her mother in Resada. As the summer passes, her much-needed student loan never materializes. Cohn takes her mind off her troubles by teaching Wilhelmina, the hornbill, to get in a crate. Cohn, who was once very disappointed that she didn't get Ebony, the raven, falls in love with the awkward bird with the bald wing.

Terri Fidone trains Kaleb to go to a mark and cush, but he is no better behaved on walks. On a Saturday in August, as Fidone walks Kaleb into Hoofstock, the camel pitches yet another fit. Fidone trips over the railroad ties along a flower bed by Schmoo's enclosure. She falls hard and cracks her head against a fence pole. The camel could easily stomp her, so Fidone bounces up as fast as she can, her crop still in her hand. Afterward, Fidone is covered with sweat. She shakes. A knot swells on her head. She sends a classmate up to Wildlife Theatre to say she can't do the rest of the show. "Still, it's been a great experience," she says of Kaleb. "I don't want to give him up. It's such an adrenaline rush."

The new vet, Cynthia Stringfield, arrives. She gives the collection a once-over. A macaw has a sour crop, a gibbon has a wound that won't heal, and Mazoe, a serval, has blood in her urine, but the collection is in good shape. Stringfield is a slim, long-limbed woman with a square jaw. At the Los Angeles Zoo, she was one of two or three vets that cared for twelve to eighteen hundred animals. She expects EATM's collection to be much more manageable. What's new to her is teaching and students—EATM students to be exact. They may prove, as Dr. Peddie has warned her, to be the most demanding species at the teaching zoo.

Beyond the zoo's fence, Dr. Peddie finds to his surprise that he takes to retirement quickly. There's no more wondering out loud about his identity. The vet finds he'd rather tinker with his boat than contemplate existential questions. He pops up to EATM off and on. In early August, he's due back on campus to receive the college's most prestigious teaching award, the Distinguished Faculty Chair. "You know, they give you an actual chair," he tells me, "a rocking chair."

Reports of the second years continue to filter in. One has gotten a job training dolphins at Sea Life Park in Hawaii. Another has gotten a boob job. Trevor Jahangard has had his tonsils out and shaved his head. After finishing up the summer camp at EATM, Mary VanHollebeke moves to Las Vegas and waits tables so she can begin paying off her student loans. After six weeks on the job, Becki Brunelli quits. The owner accused her of trying to meet movie stars. The two did not jibe, Brunelli says. She leaves LA, as well, and returns to her hometown in Northern California to be near her family. She's fallen in love and moves in with an old boyfriend. She gets a job designing Web pages. Like VanHollebeke, Brunelli puts off her dream of working with animals while she pays off her credit cards.

In a dog park along the poky Los Angeles River in Encino, Amy Mohelnitzky leans over to unhook a leash. That done, Abbey gallops across a flat expanse of grass. Her long coat blows back. Abbey stops to nose a dachshund, then she's off again. Her paws hardly seem to hit the ground as she falls in beside a mutt chasing a ball.

Abbey has taken to the life of a house dog. The pooch sleeps on a bed Mohelnitzky's mother sewed for her and sent from Wisconsin. She doesn't bark nearly as much anymore—really only here at the dog park—but who can blame her? When Mohelnitzky first brought Abbey home, she continued to feed the mutt a sloppy pile of noodles, cottage cheese, chicken broth, and rice. That wasn't working, so she tried just plain kibble with hot water. Now the dog rarely pukes. She eats nearly every meal. If Abbey hesitates to eat breakfast, Mohelnitzky calls one of her two cats over to the bowl, and that revives the dog's appetite. Like Abbey, Mohelnitzky has adjusted well to life after EATM. She works in a vet's office during the day and teaches puppy training a couple of nights a week. "It's nice to have a normal life," she says as we watch Abbey sprint, a flash of white across a seemingly endless stretch of green.

In the second week of August at about the time the pomegranates ripen on the tree near the front office, the front gate swings open and a new crop of students pours into the teaching zoo. With their arrival, the first years now officially become second years. As deeply tanned as field workers, they look around and find only three missing from their ranks. The rest have made it, even the slacker, though some have scars to show for their time here. Having survived the crucible of the last twelve months, they breathe a sigh of relief as they contemplate the riches just ahead: later mornings, field trips, and projects.

On the first day of orientation week, the new students kerthunk up the metal bleachers of Wildlife Theatre and take seats. Wilson talks to them of death. Susan Patch demonstrates how to gas a rat, and how to flick their little eyes to make sure they are dead. Another newly minted second year shows how to pull a pigeon, talking as she takes a bird in hand, leans over a trash can, and neatly twists its head off. The bleachers remain quiet, but when volunteers are requested, a good dozen hands shoot up, more than they have birds for. One newbie asks, if you don't get the head off, do you still get credit? Wilson says yes.

Just as their own second years did, the new second years agree to be kinder to their first years: they will invite them to their parties, share their doughnuts, and learn their names. They will make them feel welcome. History, however, and human behavior are against them. Few classes have jibed but it's worth a try. Maybe this underdog of a class can do it. Certainly, they've come further than the staff or they themselves expected.

The new second years look into the bleachers at the eager, fresh faces and see themselves a year ago—green, unsure, perhaps even scared. Can it be that just last August they sat there watching as Schmoo, glistening in the sun, jubilantly frolicked onstage with her trainers? They've given themselves over body and soul to the cloistered life of the teaching zoo. They've said good-bye to nearly a dozen animals. They've lived through a fire and a death. They've withstood punishing exams, gossip, and bites. They've killed a bird with their bare hands. They've learned how to walk a cougar, groom a baboon, cuddle a badger, hold a boa constrictor, and train a temperamental sea lion. Having done all that, anything seems possible.

Acknowledgments

How do you thank someone for leading the way to the most fascinating experience of your life? It was a lucky day the day I met Dr. Jim Peddie. A long list of people helped me with this project, but Dr. Peddie was the first, the one who made it all possible. This fireball of a vet was my trusty, personable guide throughout, even coming to my rescue on one occasion. He may have ruined dairy foods for me by referring to pus as "cheesy," but it was a small price to pay for his invaluable help, not to mention his company.

Likewise, I owe the outstanding staff at EATM an enormous debt: Gary and Cindy Wilson, Brenda Woodhouse, Michlyn Hines, Chuck Brinkman, Mara Rodriguez, Holly Tumas, Dorothy Belanger, Leland Shapiro, and Kris Romero. It's not easy to have a reporter knocking around asking you question after question while you leash a cougar or demonstrate how to net a monkey, but they did so with grace and good humor. They gave me open access to the school, which was key to this book. For that, I also thank the program's dean, Brenda Shubert, and Moorpark College president Eva Conrad. Former EATM staff Lynn Doria, Susan Cox, and Diane Cahill generously showed me the long view of the school. And if it weren't for Bill Brisby, EATM's founder, there would be no school for me to have written about. His big, crazy dream has made the smaller dreams of countless others come true, mine included.

Most of my time on this project was spent on the compound with the students and the animals. In fact, I often found myself hanging out at the teaching zoo long after I should have gone home. I am especially grateful to Becki Brunelli, Linda Castaneda, Susan Patch, and Trevor Jahangard, who were extraordinarily generous with their time and thoughts. I miss our cageside chats. Students Anita Wischhusen, Terri Fidone, Adam Hyde, Amber Cavett, Jenn Donovan, Carrie Hakanson, Amy Mohelnitzky, Mary VanHollebeke, Jena Anderson, Crystal Pieroni, and Chris Jenkins all lent a helping hand as well. I could go on. In short, I am forever indebted to the EATM classes of 2003 and 2004, by whose aplomb I will forever be awed.

A long list of professional animal trainers took time from their busy lives to talk to me, even let me pet an animal or two: Julie Scardina, Suzanne Morgan, and Al Kordowski at SeaWorld; Mark Forbes and Gary Mui at Birds and Animals Unlimited; Dave and Anita Jackson at Zoo to You: John Heston at the Playboy Mansion Zoo; Cathryn Hilker and Jennifer Good at the Cincinnati Zoo; Gary and Kari Johnson and their staff at Have Trunk Will Travel; Ken Ramirez at the Shedd Aquarium; Gary Priest at the San Diego Zoo and Wild Animal Park; and training revolutionary and author Karen Pryor.

As this project proved as wily as the wildest beast, my agent Jane Chelius kept me sane with her encouragement and excellent counsel. My editor, the estimable Ray Roberts, kept me on course with a potent mix of chiding and cheerleading. I suspect he was a lion tamer in a previous life. I'm also grateful to all the staff at Viking who helped make this book a reality.

I owe a debt of gratitude to one of my oldest and best friends, the talented, beautiful Lisa Stiepock, who by giving me a magazine assignment set me on the path to this book. As always, friends and family (thanks, Ma!) shored me up when the going got tough. None, however, did so as much as my soul mate, Scott Sutherland. There should be a section in heaven reserved for the spouses of authors, and if there is any justice, it will have unlimited supplies of bike magazines for my sweet, smart, handsome husband.

Lastly, how do you thank animals, the guiding spirits of this book? I owed them before I even set out on this adventure. All I can offer by way of gratitude is what any of us can, to always keep their best interests at heart. For without animals, there would be little, if any, magic in the world.